Computational Biology

Volume 26

The *Computational Biology* series publishes the very latest, high-quality research devoted to specific issues in computer-assisted analysis of biological data. The main emphasis is on current scientific developments and innovative techniques in computational biology (bioinformatics), bringing to light methods from mathematics, statistics and computer science that directly address biological problems currently under investigation.

The series offers publications that present the state-of-the-art regarding the problems in question; show computational biology/bioinformatics methods at work; and finally discuss anticipated demands regarding developments in future methodology. Titles can range from focused monographs, to undergraduate and graduate textbooks, and professional text/reference works.

More information about this series at http://www.springer.com/series/5769

Dan DeBlasio • John Kececioglu

Parameter Advising for Multiple Sequence Alignment

 Springer

Dan DeBlasio
Computational Biology Department
Carnegie Mellon University
Pittsburgh, Pennsylvania, USA

John Kececioglu
Department of Computer Science
The University of Arizona
Tucson, Arizona, USA

ISSN 1568-2684
Computational Biology
ISBN 978-3-319-87902-4   ISBN 978-3-319-64918-4   (eBook)
https://doi.org/10.1007/978-3-319-64918-4

This Springer imprint is published by Springer Nature
The registered company is Springer International Publishing AG
The registered company address is: Gewerbestrasse 11, 6330 Cham, Switzerland

To all my friends and family

DD

To Dimitri, Lorene, and Zoe

JK

Preface

While multiple sequence alignment is essential to many biological analyses, its standard formulations are all NP-complete. Due to both its practical importance and computational difficulty, a plethora of heuristic multiple sequence aligners are in use in bioinformatics. Each of these tools has a multitude of parameters which must be set, and that greatly affect the quality of the output alignment. How to choose the best parameter setting for a user's input sequences is a basic question, and most users simply rely on the aligner's default setting, which may produce a low-quality alignment of their specific sequences.

In this monograph, we present a new general approach called *parameter advising* for finding a parameter setting that produces a high-quality alignment for a given set of input sequences. In this framework, a parameter advisor is a procedure that automatically chooses a parameter setting for the aligner, and has two main ingredients: (a) the set of parameter choices considered by the advisor, and (b) an estimator of alignment accuracy used to rank alignments produced by the aligner. On coupling a parameter advisor with an aligner, once the advisor is trained in a learning phase, the user simply inputs sequences to align and receives an output alignment from the aligner, where the advisor has automatically selected the parameter setting.

The book is organized in two parts: the first lays out the foundations of parameter advising, and the second provides applications and extensions of advising. The content examines formulations of parameter advising and their computational complexity, develops methods for learning good accuracy estimators, presents approximation algorithms for finding good sets of parameter choices, and assesses software implementations of advising that perform well on real biological data. Also explored are applications of parameter advising to *adaptive local realignment*, where advising is performed on local regions of the sequences to automatically adapt to varying mutation rates; and *ensemble alignment*, where advising is applied to an ensemble of aligners to effectively yield a new aligner of higher quality than the individual aligners in the ensemble. Finally, future directions in advising research are offered.

This work arose from a series of joint research papers by the coauthors, that initiated and developed the theory and practice of parameter advising, and that formed the basis of the first author's doctoral dissertation.

Parameter advising is a general technique, with the potential to be of broad utility beyond sequence alignment. We hope this monograph encourages others to explore this fruitful area of investigation.

Dan DeBlasio   Pittsburgh, Pennsylvania
John Kececioglu   Tucson, Arizona
                                                                            October 2017

Acknowledgements

The authors gratefully acknowledge funding from the US National Science Foundation, through Grant IIS-1217886 to John Kececioglu, and by a PhD fellowship to Dan DeBlasio from the University of Arizona IGERT Program in Comparative Genomics through Grant DGE-0654435, which made this research possible. While a postdoctoral fellow at Carnegie Mellon University, Dan DeBlasio also received support from Carl Kingsford through Gordon and Betty Moore Foundation Grant GBMF4554, NSF Grant CCF-1256087, and NIH Grant R01HG007104.

Contents

Chapter 1
Introduction and Background

While multiple sequence alignment is an essential step in many biological analyses, all of the standard formulation of the problem are NP-Complete. As a consequence, many heuristic aligners are used in practice to construct multiple sequence alignments. Each of these aligners contains a large number of tunable parameters that greatly affect the accuracy of the alignment produced. In this chapter, we introduce the concept of a *parameter advisor*, which selects a setting of parameter values for a particular set of input sequences.

1.1 Multiple Sequence Alignment

The problem of aligning a set of protein sequences is a critical step for many biological analyses, including creating phylogenetic trees, predicting secondary structure, homology modeling of tertiary structure, and many others. One issue is that while we can find optimal alignments of two sequences in polynomial time [76], all of the standard formulations of the multiple sequence alignment problem are NP-complete [58, 104]. Due to the importance of *multiple sequence alignment* and its complexity, it is an active field of research.

A multiple sequence alignment of set of sequences $\{S_1, S_2 \ldots, S_k\}$ is a k by ℓ matrix of characters, where row i in the matrix contains the characters of sequence S_i, in order, possibly with gap characters inserted. The length of the alignment ℓ is at least the length of the longest sequence, so $\ell \geq \max_{1 \leq i \leq k}\{|S_i|\}$. Characters from two sequences are said to be *aligned* when they appear in the same column of the alignment. When the two aligned characters are the same, the pair is called an *identity*, otherwise it is a *mismatch*. In general, identities and mismatches are both called *substitutions*. The unweighted *edit distance* between two sequences is defined as the minimum number single-character edits to transform one sequence into the other. The edit value of an alignment is its total number

© Springer International Publishing AG 2017
D. DeBlasio, J. Kececioglu, *Parameter Advising for Multiple Sequence Alignment*,
Computational Biology 26, https://doi.org/10.1007/978-3-319-64918-4_1

of inserted gap characters and mismatches. For two sequences, you can find the optimal alignment of minimum value using dynamic programming [76]. Finding an optimal alignment of more than two sequences is NP-Complete [104]. For multiple sequence alignment, many heuristic approaches have been developed that typically use one of two common objectives. The **sum-of-pairs** score (SPS) of a multiple sequence alignment is the sum of the values of induced pairwise alignments Alternately, the **tree-alignment** objective, is the sum of pairwise alignment values align all of the branches of an input phylogenetic tree, minimized over all possible choices for ancestral sequences.

The number of alignment tools, or **aligners**, available for finding multiple sequence alignments continues to grow because of the need for high quality alignments. Many methods have been published that produce multiple sequence alignments using various heuristic strategies to deal with the problem's complexity. The most popular general method is *progressive alignment* which aligns sequences using a guide tree (a binary tree where the leaves are the input sequences [41]). Starting with two neighboring leaves a progressive aligner will optimally align these two sequences and replace the subtree that contained only these sequences by a new leaf labeled with the alignment of the two sequences. The progressive alignment method then repeats the process in a bottom-up manner aligning two of the remaining leaves (but some leaves may now contain sub-alignments). In this way a progressive aligner is only ever aligning two sequences, or alignments, to each other. This strategy has been used successfully for general alignment methods such as Clustal [98], MAFFT [54], Muscle [37, 38], Kalign [66], and Opal [106]. Additionally, progressive alignment strategies have also been successfully applied to specialized alignment tools such as those for whole genome alignment (e.g. mauve [21]) and RNA specific alignment (e.g PMFastR [31, 32] and mLoCARNA [110]). Other aligners use consistency information from a library of two-sequence alignments, such as T-Coffee [78], or collect sequence information into an HMM, as in PROMALS [83]. The majority of the studies we will present focus on the Opal aligner, the exception is Chap. 7 which examines an ensemble approach to multiple sequence alignment.

For someone who needs to use a multiple sequence alignment in biological analysis for which alignment is one of a multi-step process, choosing an aligner alone can be a daunting and somewhat time-consuming process. In addition, each tool has a set of parameters whose values can greatly impact the quality of the computed alignment. The **alignment parameters** for protein sequences typically consist of the gap-open and gap-extension penalties, as well as the choice of substitution scores for each pair of the 20 amino acids—but the available tunable parameters can differ greatly between aligners. A **parameter choice** for an aligner is an assignment of values to all of the alignment parameters. Work has been done [60] to efficiently find the optimal parameter choice for an aligner that yields the highest-accuracy alignment on average across a set of training benchmarks. This particular parameter choice would be the optimal **default parameter** choice. This choice is the one typically picked by users, as finding the best values for their particular sequences is challenging. While the default parameter choice works well on average,

```
d1gvoa   203 ... gsvenrarlvlevvdavcnewsad-RIGIRVSPigtfqnvdngpnee--adalyl--- ... 255
d2dora   141 ... ydfeatekllke-----vftfftk-PLGVKLPPyf--------------dlvhfdim ... 178
d1oyb    215 ... gsienrarftlevvdalveaighe-KVGLRLSPygvfnsmsggaetgivaqyayvage ... 272
d1o94a1  193 ... gslenrarfwletlekvkhavgsdcAIATRFGV-----------------dtvygpgq ... 234
d1ep3a   147 ... tdpevaaalvka-----ckavskv-PLYVKLSPnvt--------------divpiaka ... 185
```
(a) Higher accuracy alignment

```
d1gvoa   184 ... yl-lhqflspssnqrtdqyggsvenrarlvlevvdavcnewsad-RIGIRVSPigtfq ... 240
d2dora   159 ... kP-LGVKLPPyf--dlvhfdimaeilnqfpltYVNSV-nsig----nglfidpeaesv ... 209
d1oyb    196 ... yl-lnqfldphsntrtdeyggsienrarftlevvdalveaighe-KVGLRLSPygvfn ... 252
d1o94a1  174 ... yl-plqflnpyynkrtdkyggslenrarfwletlekvkhavgsdcAIATRF---GVdt ... 228
d1ep3a   164 ... kvPLYVKLSPnv-tdivpiakaveaagadGLTMIntl---------mgvrfdlktrqp ... 212
```
(b) Lower accuracy alignment

Fig. 1.1 (**a**) Part of an alignment of benchmark `sup_155` from the SABRE [102] suite computed by Opal [106] using non-default parameter choice (VTML200, 45, 6, 40, 40); this alignment has accuracy value 75.8%, and Facet [56] estimator value 0.423. (**b**) Alignment of the same benchmark by Opal using the *optimal* default parameter choice (BLSM62, 65, 9, 44, 43); this alignment has lower accuracy 57.3%, and lower Facet value 0.402. In both alignments, the positions that correspond to *core blocks* of the reference alignment, which should be aligned in a correct alignment, are highlighted in bold.

it can produce very low-accuracy alignments for some sequences. Figure 1.1 shows the effect of aligning the same set of five sequences under two different alignment parameter choices, one of which is the optimal default choice.

Setting the 214 parameters for the standard protein alignment model is made easier by the fact that amino acid substitution scores are well studied. Generally one of three *substitution matrix* families is used for alignment: PAM [22], BLOSUM [48], and VTML [74], but others also exist [63]. Recent work has shown that the highest-accuracy alignments are generally produced using BLOSUM and VTML matrices, so these are the only ones we consider [39].

We attempt to select a parameter choice that is best for a given input set of sequences (rather than on average) using an approach we call *parameter advising*, which we describe in the next section.

1.2 Parameter Advising

The *true accuracy* of a computed alignment is measured as the fraction of substitutions that are also present in core columns of a *reference alignment* for these sequences. (Reference alignments represent the "correct" alignment of the sequences.) These reference alignments for protein sequences are typically constructed by aligning the three-dimensional structures of the proteins. *Core columns* of this reference alignment, on which we measure accuracy, are those sections where the aligned amino acids from all of the sequences are all mutually close in three-dimensional space. Figure 1.2 shows an example of computing true accuracy for a computed alignment.

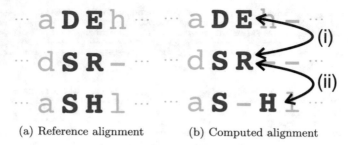

(a) Reference alignment   (b) Computed alignment

Fig. 1.2 A section of a reference and computed alignment. Accuracy of a computed alignment
(**b**) is measured with respect to a known reference alignment (**a**). We primarily use the sum-of-pairs
accuracy measure which is the fraction of aligned residues from the computed alignment recovered
in the computed alignment. In the example above the aligned residue pair (i) is correctly recovered,
while (ii) is not. This value is calculated only on core columns of an alignment (shown in red). In
the example the accuracy is 66%, because 4 of the 6 aligned residue pairs in core columns of the
reference alignment are recovered in the computed alignment.

What we have described and use throughout this book is the sum-of-pairs
definition of alignment accuracy. Another definition of multiple sequence alignment
accuracy is known as "total-column" accuracy. The total-column accuracy is the
fraction of core *columns* from the reference multiple sequence alignment that
are completely recovered in the computed alignment. For the example in Fig. 1.2
the sum-of-pairs accuracy is 66%, but the total-column accuracy is only 50%. Even
though there is only one out of place amino acid in the alignment on the right
that is from a core columns this means the whole *column* is misaligned; therefore,
only one of the two core columns is recovered in the computed alignment. While
arguments can be made for the merits of both the total-column and sum-of-pairs
accuracy measurements, the total-column measure is more sensitive to small errors
in the alignment. This is why we use the more fine-grained sum-of-pairs measure in
this book.

In the absence of a known reference alignment, we are left to estimate the
accuracy of a computed alignment. Estimating the accuracy of a computed multiple
sequence alignment (namely, how closely it agrees with the correct alignment of
its sequences), without actually knowing the correct alignment, is an important
problem. A good ***accuracy estimator*** has very broad utility: for example, from
building a meta-aligner that selects the most accurate alignment output by a
collection of aligners, to boosting the accuracy of a *single* aligner by choosing values
for the parameters of its alignment scoring function to yield the best output from that
aligner.

Given an accuracy estimator E, and a set P of parameter choices, a ***parameter
advisor*** A tries each parameter choice $p \in P$, invokes an aligner to compute an
alignment A_p using parameter choice p, and then "selects" the parameter choice p^*
that has maximum estimated accuracy $E(A_{p^*})$. Figure 1.3 shows a diagram of the
parameter advising process.

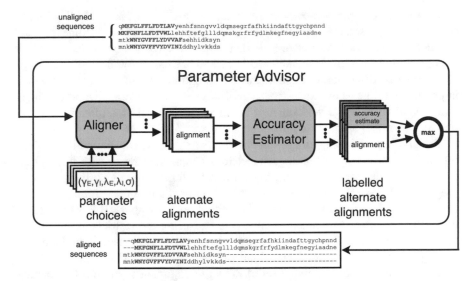

Fig. 1.3 Overview of the parameter advising process. For the Opal aligner a parameter choice consists of gap penalties $\gamma_E, \gamma_I, \lambda_E, \lambda_I$ as well as the substitution scoring matrix σ. A candidate alignment is generated for each parameter choice, so the advisor set should be small. An accuracy estimator labels each candidate alignment with an accuracy estimate. Finally, the alignment with the highest estimated accuracy is chosen by the advisor.

An advisor has two crucial elements:

(1) the **accuracy estimator**, which estimates the accuracy of a computed alignment, and which the advisor will use to choose between alternate alignments, and

(2) the **advisor set**, which is the set of parameter choices tried by the advisor to produce the alternate alignments that are chosen among.

We say that an advisor's accuracy on a set of input sequences is the true accuracy of the alignment obtained using the parameter choice selected from the advisor set with highest estimated accuracy.

We describe the recently developed **advisor estimator** called Facet (feature-based accuracy estimator) in Chaps. 2 and 3. This accuracy estimator is a linear combination of efficiently-computable feature functions. We describe the framework for the estimator and the methods for finding its coefficients in Chap. 2. The estimator coefficients are found using mathematical optimization, linear programming (LP), to identify coefficients that when used in the estimator can distinguish high accuracy alignments from low. The **feature functions** are measures of some aspect of an alignment that is easily computable and has a bounded value. The description of how to use linear programming to find an estimator as well the description of the feature functions used in Facet are described in Chap. 3.

To create a parameter advisor we also need to be able to find **advisor sets** that are of small cardinality (since the advisor will invoke the aligner for each of the parameter choices in the set) and give the best advising accuracy. An advisor set is a subset of the *parameter universe*, which is the enumeration of all possible combinations of settings for all of the parameters. Methods for finding advisor sets both for the oracle estimator (one that always returns the true accuracy of an alignment) and for a fixed estimator are described in Chap. 5. While finding optimal advisor sets is NP-complete, these methods find optimal sets of constant cardinality in polynomial time using exhaustive search. We will then describe the polynomial-time $\frac{\ell}{k}$-*approximation algorithm* for finding an advisor set of cardinality k when provided an initial optimal solution of constant size $\ell < k$.

The problem of finding an *optimal advisor* is to simultaneously find the advisor set and advisor estimator that together yield a parameter advisor with the highest possible advising accuracy. This problem can be formulated as an integer linear program (ILP), which can be restricted to find optimal advisor sets for a fixed estimator, or an optimal advisor estimator for a fixed set. Solving the ILP is intractable in practice, even for very small training sets and using the restrictions described. Finding the optimal advisor is NP-complete (see Chap. 4), as are the problems of finding an optimal advisor, and an optimal estimator (the two restrictions to the ILP).

To learn an advisor, a training set of *example* alignments whose true accuracy is known is used to find estimator coefficients and advising sets for the estimator, optimizing the advising accuracy. The training set is formed by:

(1) collecting reference alignments from standard suites of benchmark protein multiple alignments;

(2) for each such reference alignment, calling a multiple sequence aligner on the reference alignment's sequences with all parameter choices in the universe, producing a collection of alternate alignments; and

(3) labeling each such alternate alignment by its true accuracy with respect to the reference alignment for its sequences.

The benchmark suites used have reference alignments which are obtained by structural alignment of the proteins using their known three-dimensional structures. The alternate alignments together with their labeled accuracies form the *examples* in the training set. Chapter 6 describes these examples in detail and experimentally demonstrates the increase in accuracy resulting from using our new advisor.

1.3   Related Approaches

Parameter advising as described earlier can be called *a posteriori advising*: that is, advising on a parameter choice *after* seeing the resulting computed alignments. To our knowledge this is the first successful method for selecting alignment *parameter values* for a given input by choosing among computed alignments.

Work related to parameter advising can be divided into five major categories:

(i) *accuracy estimation*, which attempts to provide a score for an alignment, similar to the score produced by Facet,

(ii) *a priori advising*, which attempts to predict good parameter values for aligner from unaligned sequences as opposed to examining alignment accuracy after an alignment is generated,

(iii) *meta-alignment*, which takes the output of multiple alignment methods that are known to work well, and combines together segments of those alignments,

(iv) *column confidence scoring*, which gives a confidence score to each column in an alignment rather than the alignment as a whole, and can be used to exclude low-confidence regions of the alignment from further analysis, and

(v) *realignment methods*, sometimes also considered refinement methods, which choose sequences or sections of existing alignments and recompute an alignment in the hopes that the recomputed alignment is more accurate.

Related work from each of these categories is described below.

1.3.1   Accuracy Estimation

Several approaches have been developed for assessing the quality of a computed alignment without knowing a reference alignment for its sequences. These approaches follow two general strategies for estimating the accuracy with which a computed alignment recovers the unknown correct alignment.[1]

The first general strategy, which we call *scoring-function-based*, is to develop a new scoring function on alignments that ideally is correlated with accuracy. (See [1, 2, 77, 81, 99].) These scoring functions combine local attributes of an alignment into a score, and typically include a measure of the conservation of amino acids in alignment columns [1, 81].

The second general strategy, which we call *support-based*, is to:

(a) examine a collection of alternate alignments of the same sequences, where the collection can be generated by changing the *method* used for computing the alignment, or by changing the *input* to a method; and then

(b) measure the support for the original computed alignment among the collection of alternate alignments.

(See [61, 64, 67, 85].) In this strategy, the support for the computed alignment, which essentially measures the stability of the alignment to changes in the method or input, serves as a proxy for accuracy.

---

[1]Here *correctness* can be either in terms of the unknown *structural* alignment (as in our present work on protein sequence alignment), or the unknown *evolutionary* alignment (as in simulation studies).

1.3.1.1   Scoring-Function-Based Approaches

Approaches that assess alignment quality via a scoring function include COFFEE [77], AL2CO [81], NorMD [99], PredSP [2], and StatSigMa [86]. Several recently developed methods also consider protein tertiary (3-dimensional) structure; due to the limitations this imposes on our benchmark set we do not use them in our later analysis but they include iRMSD [7], STRIKE [59], and an LS-SVM approach [80].

COFFEE [77] uses consistency between the optimal pairwise alignments of sequences and the input multiple sequence alignment for evaluation. first a library of optimal pairwise alignments is generated by realigning each pair of sequences from the input. Then, using the matches in all these pairwise alignments the authors construct transformed substitution scores for pairs of residues[2] in the columns of the multiple alignment, where these position-dependent transformed scores are in the range $[0, 1]$. A score is then calculated by accumulating the weighted sum of scores of all induced pairwise alignments in the multiple alignment without penalizing gaps, where substitutions are evaluated using the above transformed scores. and Finally, the authors normalize the score by the weighted sum of the lengths of all induced pairwise alignments. Conceptually COFFEE is uses as the objective function for the T-Coffee aligner [78]. The "Transitive Consistency Score" (TCS) [18], from some of the same authors, uses an updated library of pairwise alignments but follows many of the same basic principals as COFFEE to construct an estimation of alignment accuracy.

AL2CO [81] uses conservation measures on alignment columns that are based on weighted frequencies of the amino acids in the column, and assesses an alignment by averaging this measure over all its columns.

NorMD [99] 5 develops an elaborate alignment scoring function that transforms standard amino acid substitution scores on pairs of aligned residues into a geometric substitution score defined on a 20-dimensional Euclidean space. A score for each column is then calculated as a weighted average of all these substitution scores in that column which is transformed through exponential decay. The final alignment score is then the sum of these transformed scores across all columns with the addition of affine gap penalties [45], and finally the Wilbur-Lipman hash scores [108] are applied as normalization factor. NorMD is used in several systems, including RASCAL [100], LEON [101], and AQUA [75].

PredSP [2] fits a beta distribution from statistics to the true accuracies associated with a sample of alignments, where the mean and variance of the distribution are transforms of a linear combination of four alignment features. The features they use are sequence percent identity, number of sequences, alignment length, and a conservation measure that is the fraction of residues in conserved columns as identified by a statistical model that takes into account amino acid background probabilities and substitution probabilities [1]. The accuracy that is predicted for

---

[2]A *residue* is a position in a protein sequence together with the amino acid at that position.

the alignment is essentially the mean of the fitted beta distribution; a predicted confidence interval on the accuracy can also be quoted from the fitted distribution.

StatSigMa [86] scores an input alignment based on a phylogenetic tree, where the tree can be given by the user or based on an alignment of a second set of sequences with the same labels. A scoring function is generated based on how well the alignment fits the tree. They then test the probability of each branch in the tree given the test alignment using Karlin-Altschul [53] alignment statistics (the same statistics used for BLAST [3] homology search). The p-value assigned to the alignment is then the maximum p-value over all branches of the tree.

iRMSD [7] uses known tertiary structure that has been assigned to all sequences in the alignment. For each pair of sequences, and for each pair of columns, they compare the distance of the pair of columns in each protein. This difference in tertiary distances is summed and weighted to generate a score for an alignment.

STRIKE [59] scoring uses a generated amino acid replacement matrix that scores based on how often two amino acids are in contact in the tertiary structure. The scoring algorithm infers the tertiary structure of a multiple sequence alignment from the known structure of a single protein in that alignment. They then examine the pairs of columns that are in contact (close in 3-D space) in the tertiary structure, and sum the STRIKE matrix score for each sequence's amino acid pairs at the columns in the alignment.

The LS-SVM approach [80] uses a similar feature-based estimator strategy to Facet. They have developed 14 feature functions for an alignment, these function each output a single numerical value and are combined to get a final accuracy estimation for an alignment. The functions used in the LS-SVM approach rely on tertiary structure, and additional information about the protein sequences obtained by querying PBD [10], PFam [42], Uniprot [6] and the Gene Ontology [GO, 14] databases—which means these databases must be available at all times. As this method makes use of tertiary structure annotations of the proteins, it severely reduces the universe of analyzable sequences. The calculated features are fed into a least-squares support vector machine (LS-SVM) that has been trained to predict accuracy.

1.3.1.2 Support-Based Approaches

Approaches that assess alignment quality in terms of support from alternate alignments include MOS [67], HoT [64], GUIDANCE [85], and PSAR [61].

MOS [67] (for "Measure of Support") takes a computed alignment together with a collection of alternate alignments of the same sequences. A score is computed by measuring the *average*, over *all* aligned residues in the input, fraction of alternate alignments that also align the residue pair. In other words, MOS measures the average support for the substitutions in the computed alignment by other alignments in the collection.

HoT [64] (for "Heads-or-Tails") considers a single alternate alignment, obtained by running the aligner that generated the computed alignment on the reverse of the sequences and reversing the resulting alignment, and reports the MOS value of the original alignment with respect to this alternate alignment.

GUIDANCE [85] assumes the computed alignment was generated by a so-called progressive aligner that uses a guide tree, and obtains alternate alignments by perturbing the guide tree and reinvoking the aligner. GUIDANCE reports the MOS value of the original alignment with respect to these alternate alignments.

PSAR [61] generates alternate alignments by probabilistically sampling pairwise alignments of each input sequence versus the pair-HMM obtained by collapsing the original alignment without the input sequence. PSAR reports the MOS value of the original alignment with respect to these alternates.

Note that in contrast to other approaches, HoT and GUIDANCE require access to the aligner that computed the alignment. They essentially measure the stability of the *aligner* to sequence reversal or guide tree alteration.

Note also that scoring-function-based approaches can estimate the accuracy of a *single* alignment, while support-based approaches inherently require a *set* of alignments and implicitly assume that a consistent alignment is the correct alignment.

1.3.2   A Priori Advising

As apposed to examining alignment accuracy after an alignment is generated, *a priori advising* attempts to make a prediction about an aligner's output from just the unaligned sequences. Such methods include AlexSys [4], PAcAlCI [79], GLProbs [114], and FSM [63].

AlexSys [4] uses a decision tree to classify the input sequences and identify which aligner should be used. At each step it tests the sequence's pairwise identity, sequence length differences, hydrophobicity characteristics, or PFam information to find relationships between sequences in the input.

PAcAlCI [79] uses similar methods to Facet and the LS-SVM approach described earlier, removing the features that rely on the alignment itself. Again features are combined using an LS-SVM. By querying multiple databases and finding similarity in tertiary structure, PAcAlCI predicts the alignment accuracy of several major alignment tools (under default parameters).

GLProbs [114] uses average pairwise percent-identity to determine which Hidden Markov Model (HMM) to use for alignment. While the actual difficulty assessment is simple, it then allows the HMM parameters to be specific to similarity of the particular sequences being aligned.

FSM [63] uses BLAST to find which family of SABmark benchmark sequences is most similar to the input sequences. It then recommends a substitution matrix that is specially tailored to these families. While BLAST is relatively fast, the applicability of this method is restricted to a narrow range of input sequences.

1.3.3   Meta-alignment

Meta-alignment uses alignments output from several aligners to construct a new alignment. Here the final alignment has aspects of the input alignments, but in contrast to other advising methods, it is not necessarily the output of any single aligner. Such methods include ComAlign [13], M-Coffee [103], Crumble and Prune[90], MergeAlign [19], and AQUA [75].

ComAlign [13] identifies paths through the standard m-dimensional dynamic programming table (which in principle would yield the optimal multiple sequence alignment of the m input sequences) that corresponds to each of the candidate input multiple sequence alignments. They then find points where these paths intersect, and construct a consensus alignment by combining the highest-scoring regions of these paths.

M-Coffee [103] uses several alignment programs to generate pairwise alignment libraries. They then use this library (rather than simply the optimal pairwise alignments) to run their T-Coffee algorithm. T-Coffee produces multiple alignments by aligning pairs of alignments to maximize the support from the library of all matching pairs of characters in each column. In this way they attempt to find the alignment with the most *support* from the other aligners.

Crumble and Prune [90] computes large-scale alignments by splitting the input sequences both vertically (by aligning subsets of sequences) and horizontally (by aligning substrings). The Crumble procedure finds similar substrings and uses any available alignment method to align these regions; then for the overlapping regions between these blocks, it realigns these intersections to generate a full alignment. The Prune procedure splits an input phylogenetic tree into subproblems with a small number of sequences; these subproblems are then replaced by their consensus sequence when parent problems are aligned, allowing large numbers of sequences to be aligned. This method aims to reduce the computational resources needed to align large inputs, as opposed to increasing multiple sequence alignment accuracy.

MergeAlign [19] generates a weighted directed acyclic graph (DAG), where each vertex represents a column in one of the input alignments, and an edge represents a transition from one column to its neighbor (the column directly to the right) in the same alignment. The weight of each edge is the number of alignments that have that transition. The consensus alignment is constructed as the maximum-weight single-source/single-sink path in the DAG.

AQUA [75] chooses between an alignment computed by Muscle [38] or MAFFT [55], based on their NorMD [99] score. Chapter 6 shows that for the task of choosing the more accurate alignment, the NorMD score used by AQUA is much weaker than the Facet estimator used here. AQUA can also be used as a ***meta-aligner*** because it chooses between the outputs of *multiple aligners*, rather than two parameter choices for a *singe aligner*. Chapter 7 gives results on using Facet in the context of meta-alignment.

1.3.4 Column Confidence Scoring

In addition to scoring whole alignments, work has been done to identify poorly-aligned regions of alignments. This can help biologists to find unreliable homology in an alignment to ignore for further analysis, as in programs like GBLOCKS [17] and ALISCORE [72]. Many of the accuracy estimators discussed earlier also provide column level scoring, such as TCS.

GBLOCKS [17] identifies columns that are conserved and surrounded by other conserved regions, using only column percent-identity. Columns that contain gaps are eliminated, as well as runs of conservation that do not meet length cutoffs.

ALISCORE [72] uses a sliding window across all pairwise alignments, and determines if the window is statistically different from two random sequences. This score is evaluated on a column by counting the number of windows containing it that are significantly non-random.

Recently, Ren [88] developed a new method to classify columns with an SVM. This method uses five features of an alignment that are passed into the SVM to classify whether or not the column should be used for further analysis. Their study focuses mainly on using alignments for phylogenetic reconstruction.

1.3.5 Realignment Methods

Methods that partition a set of sequences to align or realign them can be divided in two categories, based on the type of partition. *Vertical* realigners cut the input sequences into substrings, and once these shorter substrings are realigned, they stitch their alignments together. *Horizontal* realigners split an alignment into groups of whole sequences, which are then merged together by realigning between groups, possibly using each group's induced subalignment.

Crumble and Prune [90] is a pair of algorithms for performing both vertical (Crumble) and horizontal (Prune) splits on an input set of sequences. During the Crumble stage, a set of constraints is found that anchor the input sequences together, and the substrings or blocks between these anchor points are aligned. Once the disjoint blocks of the sequences are aligned, they are then fused by aligning their overlapping anchor regions. During the Prune stage, smaller groups of sequences are aligned that correspond to subtrees of the progressive aligner's guide tree. The subset of sequences in a subtree is then replaced by their alignment's consensus sequence in the remaining steps of progressive alignment. The original subalignments of the groups are finally reinserted to form the output alignment. Replacing a group of sequences by their consensus sequence during alignment reduces the number of sequences that are aligned at any one time. The objective for splitting sequences both vertically and horizontally within Crumble and Prune is to reduce time and space usage to make feasible the alignment of large numbers of long sequences, rather than to improve alignment accuracy.

An early example of horizontal realignment is ReAligner [5] which improves DNA sequence assembly by removing and then realigning sequencing reads. If a read is initially misaligned in the assembly it may be corrected when the read is removed and realigned. This realignment process is repeated over all reads to continually refine the assembly.

Gotoh [46] presented several horizontal methods for heuristically aligning two multiple sequence alignments, which he called "group-to-group" alignment. This could be used for alignment construction in a progressive aligner, proceeding bottom-up over the guide tree and applying group-to-group alignment at each node, or for polishing an existing alignment by assigning sequences to two groups and using it to realign the groups.

The standard aligners Muscle [37, 38], MAFFT [55], and ProbCons [34] also include a polishing step that performs horizontal realignment similar to Gotoh.

AlignAlign [58] is a horizontal method that implements an exact algorithm for optimally aligning two multiple sequence alignments under the sum-of-pairs scoring function with affine gap costs. This optimal group-to-group alignment algorithm, used for both alignment construction and alignment polishing, forms the basis of the Opal aligner [106].

1.4   Background on Protein Structure

This monograph focuses specifically on parameter advising for multiple sequence alignment of *proteins*. Because of this focus on proteins, which are folded in the cell into three-dimensional structures to perform their functions, many of the methods described in later chapters (and some of the methods mentioned in the previous section) make use of this folding to predict the accuracy of alignments. This section briefly reviews the concepts surrounding protein structure as used in the rest of the book.

Multiple sequence alignment benchmarks of protein sequences are normally constructed by aligning the three-dimensional structures, sometimes referred to as *tertiary structure*, of the folded proteins. Amino acids that are close in space are considered aligned, and those that are simultaneously very close in all sequences are labeled as *core columns* of the alignment. The amino acid sequence of a protein is referred to as the *primary structure*. The *secondary structure* of a protein is an intermediate between primary and tertiary structure. Secondary structure labels each amino acid as being in one of three structure classes: *α-helix*, *β-sheet*, or *coil* (also called other, representing no secondary structure). These structural classes tend to be conserved when the function of related proteins is conserved.

The tertiary structure of the proteins in a set of input sequences is not normally known, as it usually requires examining the crystalline structure of the protein, which is slow and costly. Instead we use secondary structure in our accuracy estimator. To predict secondary structure, the PSIPRED [51] tool is used to predict structure for all of the experimentation shown in later chapters. The output of PSIPRED is not only a label from the three secondary structure classes for each amino acid in the sequence, but also a confidence that the position in each sequence is in each structure state. For most applications in the rest of this book the confidences are normalized such that for any amino acid the sum of the confidences for all three structure types sums to 1.

PSIPRED can make predictions using either the amino acid sequence alone, or by searching through a database of protein domains to find similar sequences using BLAST [3]. The accuracy of PSIPRED is increased substantially when using a BLAST search, so all of our results shown later are with the version of PSIPRED that searches through the UniRef90 [96] database of protein domains (which is a non-redundant set of domains from the UniProt database see [97]) filtered using the pfilt program provided with PSIPRED.

1.5   Overview

Chapter 2 next describes an approach to estimating alignment accuracy as a linear combination of feature functions. It also describes how to find the coefficients for such an estimator.

Chapter 3 describes the Facet estimator (short for "*f*eature-based *ac*curacy *es*timator"). In particular, the chapter describes the efficiently-computable feature functions used in Facet.

Chapter 4 defines the problem of finding an optimal advisor (simultaneously finding both the advisor set and the estimator coefficients). We also consider restrictions to finding just an optimal advisor set, or optimal advisor coefficients. We show that all of these problems are NP-complete.

Chapter 5 details methods for finding advisor sets for a fixed estimator. While finding an optimal advisor set is NP-Complete, we present an efficient approximation algorithm for finding near-optimal sets that performs well in practice. The chapter also describes an integer linear program for finding an optimal advisor (which at present cannot be solved to optimality in practice).

Chapter 6 provides experimental results for parameter advising, and discusses the approach used to assess the effectiveness of advising.

Chapter 7 expands the universe of parameter choices in advising to include not only the settings of the alignment parameters, but also the choice of the aligner itself, which we call *aligner advising*. This yields the first true ***ensemble aligner***. We also compare the accuracy of the alignments produced by the ensemble aligner to those obtained using a parameter advisor with a fixed aligner.

Chapter 8 presents an approach called ***adaptive local realignment***, which computes alignments that can use different parameter choices in different regions of the alignment. Since regions of a protein have distinct mutation rates, using different parameter choices across an alignment can be necessary.

Chapter 9 describes an approach to predicting how much of a column in a computed alignment comes from core columns of an unknown reference alignment, using a variant of nearest-neighbor classification. Since true accuracy is only measured on core columns, inferring such columns can boost the accuracy of our advisor.

Finally, Chap. 10 offers directions for future research.

Part I
Foundations of Parameter Advising

Chapter 2
Alignment Accuracy Estimation

As we discussed earlier, the accuracy of a multiple sequence alignment is commonly measured as the fraction of aligned residues from the core columns of a known reference alignment that are also aligned in the computed alignment. Usually this reference alignment is unavailable, in which case we can only *estimate* the accuracy. In this chapter we will describe a reference-free approach that estimates accuracy as a linear combination of bounded feature functions of an alignment. In the following sections we first introduce the framework for accuracy estimation and show that all higher-order polynomial estimators can be reduced to a linear estimator. We go on to give several approaches for learning the coefficients of the estimator function through mathematical optimization.

2.1 Constructing Estimators from Feature Functions

Without knowing a reference alignment that establishes the ground truth against which the true accuracy of an alignment is measured, we are left with only being able to estimate the accuracy of an alignment. The approach we describe here to obtain an estimator for alignment accuracy is to (a) identify multiple *features* of an alignment that tend to be correlated with accuracy, and (b) combine these features into a single accuracy estimate. Each feature, as well as the final accuracy estimator, is a real-valued function of an alignment.

The simplest estimator is a linear combination of feature functions, where features are weighted by coefficients. These coefficients can be learned by training the estimator on example alignments whose true accuracy is known. This training process will result in a fixed coefficient or weight for each feature. Alignment

---

Adapted from publications [33, 56].

D. DeBlasio, J. Kececioglu, *Parameter Advising for Multiple Sequence Alignment*,
Computational Biology 26, https://doi.org/10.1007/978-3-319-64918-4_2

accuracy is usually represented by a value in the range $[0, 1]$, with 1 corresponding to perfect accuracy. Consequently, the value of the estimator on an alignment should be bounded, no matter how long the alignment or how many sequences it aligns. For boundedness to hold when using fixed feature weights, the feature functions themselves must also be bounded. Hence, it is assumed that the feature functions also have the range $[0, 1]$. (The particular features used are presented in Chap. 3.) The estimator can then be guaranteed to have the range $[0, 1]$ by ensuring that the coefficients found by the training process yield a convex combination of features. In practice, not all the features naturally span the entire range $[0, 1]$, so the convex combination condition can be relaxed such that the only requirement is that the estimator value fall in the range $[0, 1]$ on all training examples.

In general, lets consider estimators that are polynomial functions of alignment features. More precisely, suppose the features considered for alignments A are measured by the k feature functions $f_i(A)$ for $1 \leq i \leq k$. Then the accuracy estimator $E(A)$ is a polynomial in the k variables $f_i(A)$. For example, for a degree-2 polynomial,

$$E(A) := a_0 + \sum_{1 \leq i \leq k} a_i f_i(A) + \sum_{1 \leq i,j \leq k} a_{ij} f_i(A) f_j(A).$$

For a polynomial of degree d, the *accuracy estimator $E(A)$* has the general form,

$$E(A) := \sum_{\substack{p_1,\ldots,p_k \in \mathcal{Z}^+ \\ p_1 + \cdots + p_k \leq d}} a_{p_1,\ldots,p_k} \prod_{1 \leq i \leq k} \left(f_i(A) \right)^{p_i},$$

where $\mathcal{Z}^+$ denotes the nonnegative integers, and the coefficients on the terms of the polynomial are given by the $a_{p_1,\ldots,p_k}$. In this summation, there are k index variables p_i, and each possible *assignment* of nonnegative integers to the p_i that satisfies $\sum_i p_i \leq d$ specifies one *term* of the summation, and hence the powers for one term of the polynomial.

Encoding Polynomial Estimators

Learning an estimator from example alignments, as discussed in Sect. 2.2, corresponds to determining the coefficients for its terms. Optimal coefficients can be learned efficiently that minimize the error between the estimate $E(A)$ and the actual accuracy of alignment A on a set of training examples, even for estimators that are polynomials of *arbitrary* degree d. This can be done for arbitrary degree essentially because such an estimator can always be reduced to the linear case by a change of feature functions, as follows. For *each* term in the degree-d estimator, where the term is specified by the powers p_i of the f_i, define a *new* feature function

$$g_j(A) \;:=\; \prod_{1 \le i \le k} \left(f_i(A) \right)^{p_i},$$

that has an associated coefficient $c_j := a_{p_1,\ldots,p_k}$. Then in terms of the new feature functions g_j, the original degree-d estimator is equivalent to the linear estimator

$$E(A) \;=\; c_0 + \sum_{1 \le j < t} c_j \, g_j(A),$$

where t is the number of terms in the original polynomial. For a degree-d estimator with k original feature functions, the number of coefficients t in the linearized estimator is at least $\mathcal{P}(d, k)$, the number of integer partitions of d with k parts. This number of coefficients grows very fast with d, so overfitting can become an issue when learning a high-degree estimator. (Even a cubic estimator on ten features already has 286 coefficients.) The experiments in later chapters will focus only on *linear* estimators.

The coefficients of the estimator polynomial are found by mathematical optimization which is described next.

2.2 Learning the Estimator from Examples

Section 1.2 described the set of *examples* as: benchmark sequences that have been aligned under various parameter choices by the aligner, and whose alignment are labeled with their true accuracy. In addition, each example is labeled with the feature function values for each of these examples. These examples can then be used, with their associated accuracy and feature values, to find coefficients that fit the accuracy estimator to true accuracy by either of the two techniques describe below.

2.2.1 Fitting to Accuracy Values

A natural criterion for fitting the estimator is to minimize the error on the example alignments between the estimator and the true accuracy value. For alignment A in the training set $\mathcal{S}$, let $E_c(A)$ be its estimated accuracy where vector $c = (c_0, \ldots, c_{t-1})$ specifies the values for the coefficients of the estimator polynomial, and let $F(A)$ be the *true accuracy* of example A.

Formally, minimizing the weighted error between estimated accuracy and true accuracy yields estimator $E^* := E_{c^*}$ with coefficient vector

$$c^* \;:=\; \operatorname*{argmin}_{c \in \mathcal{R}^t} \sum_{A \in \mathcal{S}} w_A \left| E_c(A) - F(A) \right|^p,$$

where power p controls the degree to which large accuracy errors are penalized. Weights w_A correct for sampling bias among the examples, as explained below.

When $p = 2$, this corresponds to minimizing the L_2 norm between the estimator and the true accuracies. The absolute value in the objective function may be removed, and the formulation becomes a *quadratic programming problem* in variables c, which can be efficiently solved. (Note that E_c is linear in c.) When $p = 1$, the formulation corresponds to minimizing the L_1 norm. This is less sensitive to outliers than the L_2 norm, which can be advantageous when the underlying features are noisy. Minimizing the L_1 norm can be reduced to a *linear programming problem* as follows. In addition to variables c, let e be a second vector of variables with an entry e_A for each example $A \in S$ to capture the absolute value in the L_1 norm, along with the inequalities,

$$e_A \geq E_c(A) - F(A),$$

$$e_A \geq F(A) - E_c(A),$$

which are linear in variables c and e. The objective is to then minimize the linear function

$$\sum_{A \in S} w_A \, e_A.$$

For n examples, the linear program has $n + t$ variables and $O(n)$ inequalities, which is solvable even for very large numbers of examples.

If the feature functions all have range $[0, 1]$, the resulting estimator E^* can be ensured to also have range $[0, 1]$ by adding to the linear inequalities,

$$c_i \geq 0,$$

$$\sum_{0 \leq i < t} c_i \leq 1.$$

But as mentioned earlier, it may be useful to not restrict the coefficients to be a convex combination because while the features are bounded, they may have an inconsistent range of probable values. This set of equations can instead be replaced by the following inequalities for each training example A that ensure E^* has range $[0, 1]$.

$$E_c(A) \geq 0,$$

$$E_c(A) \leq 1.$$

The weights w_A on examples aid in finding an estimator that is good across all accuracies. In the suites of protein alignment benchmarks that are commonly available, a predominance of the benchmarks consist of sequences that are easily alignable, meaning that standard aligners find high-accuracy alignments for these benchmarks.[1] In this situation, when training set $\mathcal{S}$ is generated as described earlier, most examples have high accuracy, with relatively few at moderate to low accuracies. Without weights on examples, the resulting estimator E^* is strongly biased towards optimizing the fit for high accuracy alignments, at the expense of a poor fit at lower accuracies. To prevent this, the examples in $\mathcal{S}$ are binned by their true accuracy, where $\mathcal{B}(A) \subseteq \mathcal{S}$ is the set of alignments falling in the bin for example A, and a weight is assigned to the error term for A as $w_A := 1/|\mathcal{B}(A)|$. (In the experiments we show later, 10 bins equally spaced bins are formed at 10% increments in accuracy.) In the objective function this weights *bins* uniformly (rather than weighting *examples* uniformly) and weights the error equally across the full range of accuracies.

2.2.2 Fitting to Accuracy Differences

Many applications of an accuracy estimator E will use it to choose from a set of alignments the one that is estimated to be most accurate. (This occurs, for instance, in parameter advising as discussed briefly in Chap. 1.) In such applications, the estimator is effectively ranking alignments, and all that is needed is for the estimator to be *monotonic* in true accuracy. Accordingly, rather than trying to fit the estimator to match accuracy *values*, it would be more useful to instead fit it so that *differences* in accuracy are reflected by at least as large differences in the estimator. This fitting to differences is less constraining than fitting to values, and hence might be better achieved.

More precisely, select a set $\mathcal{P} \subseteq \mathcal{S}^2$ of ordered pairs of example alignments, where every pair $(A, B) \in \mathcal{P}$ satisfies $F(A) < F(B)$. Set $\mathcal{P}$ holds pairs of examples on which accuracy F increases and therefore similar behavior from our estimator E is desired. (Later we discuss how to select a small set $\mathcal{P}$ of important pairs.) If estimator E increases at least as much as accuracy F on a pair in $\mathcal{P}$, this is a success, and if it increases less than F, the amount it falls short is considered an error, which will be minimized. Notice this tries to match large accuracy increases, and penalizes less for not matching small increases.

Formally, fitting to differences is formulated as finding the optimal estimator $E^* := E_{c^*}$ given by coefficients

$$ c^* := \operatorname*{argmin}_{c \in \mathcal{R}^I} \sum_{(A,B) \in \mathcal{P}} w_{AB} \left(\max \left\{ \left(F(B) - F(A)\right) - \left(E_c(B) - E_c(A)\right), \, 0 \right\} \right)^p, $$

---

[1]This is mainly a consequence of the fact that proteins for which reliable structural reference alignments are available tend to be closely related, and hence easier to align. It does not mean that typical biological inputs are easy.

where w_{AB} weights the error term for a pair. When power p is 1 or 2, we can reduce this optimization problem to a linear or quadratic program as follows. First define a vector of variables e with an entry e_{AB} for each pair $(A, B) \in \mathcal{P}$, along with the inequalities,

$$e_{AB} \geq 0,$$

$$e_{AB} \geq \big(F(B) - F(A)\big) - \big(E_c(B) - E_c(A)\big),$$

which are linear in variables c and e. The objective will then be to minimize the function,

$$\sum_{(A,B) \in \mathcal{P}} w_{AB} \, (e_{AB})^p,$$

which is linear or quadratic in the variables for $p = 1$ or 2.

For a set $\mathcal{P}$ of m pairs, these programs have $m + t$ variables and m inequalities, where $m = O(n^2)$ in terms of the number of examples n. For the programs to be manageable for large n, set $\mathcal{P}$ must be quite sparse.

To select a sparse set $\mathcal{P}$ of important pairs use one of two methods: threshold-minimum accuracy difference pairs, or distributed-example pairs. Recall that the training set $\mathcal{S}$ of examples consists of alternate alignments of the sequences in benchmark reference alignments, where the alternates are generated by aligning the benchmark under a constant number of different parameter choices.

Threshold-Difference Pairs

While it is desirable to have an accuracy estimator that matches the difference in true accuracy between *any* two alignments, in parameter advising the only concern is choosing among alignments over the *same* sets of sequences. With threshold-difference pairs, $\mathcal{P}$ only includes pairs of alignments (A, B) of the *same benchmark*. In particular, it includes all such pairs where $F(A) - F(B) \geq \epsilon$. Here $\epsilon > 0$ is a tunable threshold; if the difference in accuracy is smaller than this threshold, it is excluded from training, as its effect on the parameter advisor is minimal, and it makes the linear or quadratic problem much harder to solve. As ϵ approaches 0, the better the estimator will be at distinguishing small differences, but more constraints will be included in the program increasing the running time of the solver. For example pairs under this model, we set the weight w_{AB} to be $\frac{1}{|\mathcal{B}(C)|}$, where $\mathcal{B}$ gives the corresponding bin for benchmarks aligned under the default parameter settings, and C is the alignment under the default parameter settings of the benchmark sequences that A and B are aligning.

Notice that the threshold-difference method for selecting $\mathcal{P}$ could choose $O(p^2)$ pairs of alignments for each benchmark when $\mathcal{S}$ contains p alternate alignments for that benchmark. This bound is only reached if all of the alignments are very different (though in this case the pairs make more of a contribution to the advisor accuracy), but would select many fewer pairs if the alignments are all similar. To further constrain the size of $\mathcal{P}$ a second tunable parameter k is introduced and is the maximum number of pairs of alternate alignments for each benchmark that is allowed to be included.

Therefore the size of $\mathcal{P}$ using the threshold-difference method is bounded by $k\,b$ where b is the number of benchmarks in the training set, but with large value of ϵ may be much smaller.

Distributed-Example Pairs

An estimator that is designed for parameter advising should rank the highest accuracy alternate alignment for a benchmark above the other alternates for that benchmark. Consequently, for each benchmark the distributed examples method selects for $\mathcal{P}$ its highest-accuracy alternate, paired with its other alternates for which their difference in accuracy is at least ϵ, where ϵ is a tunable threshold. (Notice this picks $O(n)$ pairs on the n examples.) For the estimator to generalize beyond the training set, it helps to also properly rank alignments between benchmarks. To include some pairs between benchmarks, the minimum, maximum, and median accuracy alignments are chosen for each benchmark, and form one list L of all these chosen alignments, ordered by increasing accuracy. Then for each alignment A in L, while making a scan of L to the right select the first k pairs (A, B) for which $F(B) \geq F(A) + i\,\delta$ where $i = 1, \ldots, k$, and for which B is from a different benchmark than A. While the constants $\epsilon \geq 0$, $\delta \geq 0$, and $k \geq 1$ control the specific pairs that this procedure selects for $\mathcal{P}$, it always selects $O(n)$ pairs on the n examples.

Weighting distributed-example pairs When fitting to accuracy differences, weights are assigned to the error terms to correct for sampling bias within $\mathcal{P}$, but now these terms are associated with pairs of alignment. The pairs should be weighted such that optimization treats the entire accuracy range equally, so the fitted estimator performs well at all accuracies. As when fitting to accuracy values, the example alignments in S are partitioned into bins $\mathcal{B}_1, \ldots, \mathcal{B}_k$ according to their true accuracy. To model equal weighting of accuracy bins by pairs, consider a pair $(A, B) \in \mathcal{P}$ to have half its weight w_{AB} on the bin containing A, and half on the bin containing B. (So in this model, a pair (A, B) with both ends A, B in the same bin $\mathcal{B}$, places all its weight w_{AB} on $\mathcal{B}$.) Then the problem is to find weights $w_{AB} > 0$ that, for all bins $\mathcal{B}$, satisfy

$$\sum_{(A,B) \in \mathcal{P} \,:\, A \in \mathcal{B}} \tfrac{1}{2} w_{AB} \;+\; \sum_{(A,B) \in \mathcal{P} \,:\, B \in \mathcal{B}} \tfrac{1}{2} w_{AB} \;=\; 1.$$

In other words, the pairs should weight bins uniformly.

A collection of weights w_{AB} are *balanced* if they satisfy the above property on all bins $\mathcal{B}$. While balanced weights do not always exist in general, an easily-satisfied condition can be identified that guarantees they do exist, and in this case find balanced weights by the following graph algorithm.

Construct an undirected graph G whose vertices are the bins B_i and whose edges (i, j) go between bins B_i, B_j that have an alignment pair (A, B) in $\mathcal{P}$ with $A \in B_i$ and $B \in B_j$. (Notice G has self-loops when pairs have both alignments in the same bin.) Our algorithm first computes weights ω_{ij} on the edges (i, j) in G, and then assigns weights to pairs (A, B) in $\mathcal{P}$ by setting $w_{AB} := 2\,\omega_{ij}/c_{ij}$, where bins B_i, B_j contain alignments A, B, and c_{ij} counts the number of pairs in $\mathcal{P}$ between bins B_i and B_j. (The factor of 2 is due to a pair only contributing weight $\frac{1}{2}w_{AB}$ to a bin.) A consequence is that all pairs (A, B) that go between the same bins get the same weight w_{AB}.

During the algorithm, an edge (i, j) in G is said to be *labeled* if its weight ω_{ij} has been determined; otherwise it is *unlabeled*. Call the *degree* of a vertex i the total number of endpoints of edges in G that touch i, where a self-loop contributes two endpoints to the degree. Initially all edges of G are unlabeled. The algorithm sorts the vertices of G in order of nonincreasing degree, and then processes the vertices from highest degree on down.

In the general step, the algorithm processes vertex i as follows. It accumulates w, the sum of the weights ω_{ij} of all *labeled* edges that touch i; counts u, the number of *unlabeled* edges touching i that are not a self-loop; and determines d, the degree of i. To the unlabeled edges (i, j) touching i, the algorithm assigns weight $\omega_{ij} := 1/d$ if the edge is not a self-loop, and weight $\omega_{ii} := \frac{1}{2}(1 - w - \frac{u}{d})$ otherwise.

This algorithm assigns *balanced weights* if in graph G, every bin has a self-loop, as stated in the following theorem.

Theorem 2.1 (Finding Balanced Weights) *Suppose every bin B has some pair (A, B) in $\mathcal{P}$ with both alignments A, B in B. Then the above graph algorithm finds balanced weights.*

Proof. We will show that: (a) for every edge (i, j) in G, its assigned weight satisfies $\omega_{ij} > 0$; and (b) for every vertex i, the weights assigned to its incident edges (i, j) satisfy

$$\sum_{(i,j)\,:\,j \neq i} \omega_{ij} + 2\,\omega_{ii} = 1.$$

From properties (a) and (b) it follows that the resulting weights w_{AB} are balanced.

The key observation is that when processing a vertex i of degree d, the edges touching i that are already *labeled* will have been assigned a weight $\omega_{ij} \leq 1/d$, since the other endpoint j must have degree at least d (as vertices are processed from highest degree on down). Unlabeled edges touching i, other than a self-loop, get assigned weight $\omega_{ij} = 1/d > 0$. When assigning weight ω_{ii} for the unlabeled self-loop, the total weight w of incident labeled edges satisfies $w \leq (d-u-2)/d$, by the key observation above and the fact that vertex i always has a self-loop which contributes 2 to its degree. This inequality in turn implies $\omega_{ii} \geq 1/d > 0$. Thus property (a) holds.

Furthermore, twice the weight ω_{ii} assigned to the self-loop takes up the slack between 1 and the weights of all other incident edges, so property (b) holds as well. □

Regarding the condition in Theorem 2.1, if there are bins without self-loops, balanced weights do not necessarily exist. The smallest such instance is when G is a path of length 2.

Notice that the condition in Theorem 2.1 can be ensured to hold if every bin has at least two example alignments: simply add a pair (A, B) to $\mathcal{P}$ where both alignments are in the bin, if the procedure for selecting a sparse $\mathcal{P}$ did not already. When the training set S of example alignments is sufficiently large compared to the number of bins (which is within our control), every bin is likely to have at least two examples. So Theorem 2.1 essentially guarantees that in practice the estimator can be fit to examples chosen using the distributed-example pairs method using balanced weights.

For k bins and m pairs, the pair-weighting algorithm can be implemented to run in $O(k + m)$ time, using radix sort to map pairs in $\mathcal{P}$ to edges in G, and counting sort to order the vertices of G by degree.

Summary

In this chapter, we have described an accuracy estimator that is a linear combination of feature functions, and provided two approaches to learning the coefficients of this estimator. Chapter 3 next gives details of the specific features that along with this framework make up the Facet accuracy estimator. Results on using Facet with the feature functions described in the next chapter are presented in Chap. 6.

Chapter 3
The Facet Accuracy Estimator

In Chap. 2, we described a general framework for creating an alignment accuracy estimator that is a linear combination of feature functions, and two methods for learning the coefficients of such an estimator. In this chapter, we explore the feature functions used in the Facet accuracy estimator. Some of the features considered are standard metrics that are common for measuring multiple sequence alignment quality, such as amino acid percent identity and gap extension density, but many of the most reliable features are novel. The strongest feature functions tend to use predicted secondary structure. We describe in detail the most accurate and novel features: Secondary Structure Blockiness, and Secondary Structure Agreement.

3.1 Feature Functions of an Alignment

Given the framework presented in the previous chapter, what makes an accuracy estimator unique are the features it uses. The feature-based accuracy estimator we call Facet is one instantiation of a *scoring-function-based* accuracy estimator that uses this framework. What sets Facet apart are *novel feature functions* that measure non-local properties of an alignment, and have stronger correlation with true accuracy (such as Secondary Structure Blockiness, described in Sect. 3.2).

In addition to the features currently available within Facet, this approach allows for the incorporation of new feature functions into the estimator, and is easily tailored to a particular class of alignments by choosing appropriate features and performing regression.

Adapted from publications [33, 56].

© Springer International Publishing AG 2017
D. DeBlasio, J. Kececioglu, *Parameter Advising for Multiple Sequence Alignment*,
Computational Biology 26, https://doi.org/10.1007/978-3-319-64918-4_3

Compared to support-based approaches, Facet does not degrade on difficult alignment instances, where for parameter advising, good accuracy estimation can have the greatest impact. As shown in the advising experiments in Chap. 6, support-based approaches lose the ability to detect accurate alignments of hard-to-align sequences, since for such sequences most alternate alignments are poor and lend little support to the alignment that is actually most accurate.

The remainder of this chapter describes twelve feature functions on alignments, the majority of which are novel to Facet. All are efficiently computable, so the resulting estimator is fast to evaluate. The strongest feature functions make use of predicted *secondary structure* (which is not surprising, given that protein sequence alignments are often surrogates for structural alignments). Background on protein secondary structure, and how we predict it for new proteins, is described in Sect. 1.4.

Another aspect of some of the best alignment features is that they tend to use *non-local information*. This is in contrast to standard ways of scoring sequence alignments, such as with amino acid substitution scores or gap open and extension penalties, which are often a function of a single alignment column or two adjacent columns (as is necessary for efficient dynamic programming algorithms). While a good accuracy estimator would make an ideal scoring function for *constructing* a sequence alignment, computing an optimal alignment under such a nonlocal scoring function seems prohibitive (especially since multiple alignment is already NP-complete for the current highly-local scoring functions). Nevertheless, given that our estimator can be efficiently evaluated on any constructed alignment, it is well suited for *selecting* a sequence alignment from among several alternate alignments, as we discuss in Chap. 6 in the context of parameter advising (while later chapters further consider the contexts of ensemble alignment and adaptive local realignment).

Key properties of a good feature function are that it should: (a) measure an attribute that discriminates high-accuracy alignments from others, (b) be efficiently computable, and (c) be bounded in value (as discussed at the beginning of Chap. 2). Bounded functions can be normalized, and we scale all our feature functions to the range [0, 1]. We also intend our feature functions to be positively correlated with alignment accuracy.

The following are the alignment feature functions considered by Facet. We highlight the first function as it is the most novel, one of the strongest, and is the most challenging to compute.

3.2 Secondary Structure Blockiness

The reference alignments in the most reliable suites of protein alignment benchmarks are computed by structural alignment of the known three-dimensional structures of the proteins. The so-called *core blocks* of these reference alignments, which are the columns in the reference to which an alternate alignment is compared when measuring its true accuracy, are typically defined as the regions of the structural alignment in which the residues of the different proteins are all within a small distance threshold of each other in the superimposed structures. These regions of structural agreement are usually in the embedded core of the folded proteins, and the secondary structure of the core usually consists of α-helices and β-strands. (Details of *secondary structure* and its representation can be found in Sect. 1.4.) As a consequence, in the reference sequence alignment, the sequences in a core block often share the same secondary structure, and the type of this structure is usually α-helix or β-strand.

The degree to which a multiple alignment displays this pattern of structure is measured by a feature called *Secondary Structure Blockiness*. As mentioned earlier this feature, as with many others, utilizes the secondary structure of each protein, given by a standard prediction tool such as PSIPRED [51]. Then in multiple sequence alignment A and for given integers $k, \ell > 1$, define a *secondary structure block* $\mathcal{B}$ to be:

(i) a contiguous interval of at least ℓ columns of A, together with

(ii) a subset of at least k sequences in A, such that on all columns in this interval, in all sequences in this subset, all the entries in these columns for these sequences have the same predicted secondary structure type, and this shared type is all α-helix or all β-strand.

$\mathcal{B}$ is called an α-block or a β-block according to the common type of its entries. Parameter ℓ, which controls the minimum width of a block, relates to the minimum length of α-helices and β-strands; the definition can be extended to use different values ℓ_α and ℓ_β for α- and β-blocks.

A *packing* for alignment A is a set $\mathcal{P} = \{\mathcal{B}_1, \ldots, \mathcal{B}_b\}$ of secondary structure blocks of A, such that the column intervals of the $\mathcal{B}_i \in \mathcal{P}$ are all disjoint. (In other words, in a packing, each column of A is in at most one block. The sequence subsets for the blocks can differ arbitrarily.) The *value* of a block is the total number of residue pairs (or equivalently, substitutions) in its columns; the value of a packing is the sum of the values of its blocks.

Finally, the *blockiness* of an alignment A is the maximum value of any packing for A, divided by the total number of residue pairs in the columns of A. In other words, Secondary Structure Blockiness measures the fraction of substitutions in A that are in an optimal packing of α- or β-blocks.

At first glance measuring blockiness might seem hard (since optimal packing problems are often computationally intractable), yet surprisingly it can actually be computed in *linear time* in the size of the alignment, as the following theorem states. The main idea is that evaluating blockiness can be reduced to solving a longest path problem on a directed acyclic graph of linear size.

Theorem 3.1 (Evaluating Blockiness) *Given a multiple alignment A of m protein sequences and n columns, where the sequences are annotated with predicted secondary structure, the blockiness of A can be computed in $O(mn)$ time.*

Proof. The key is to not enumerate subsets of sequences in A when considering blocks for packings, and instead enumerate intervals of columns of A. Given a candidate column interval I for a block $\mathcal{B}$, since there are only two possibilities for the secondary structure type s of $\mathcal{B}$, and the sequences in $\mathcal{B}$ must have type s

across I, considering all possible subsets of sequences is avoided. To maximize the *value* of $\mathcal{B}$, first collect all sequences in A that have type α across I (if any), all sequences that have type β across I, and keep whichever subset has more sequences.

Following this idea, given alignment A, form an edge-weighted, directed graph G that has a vertex for every column of A, plus an artificial *sink* vertex, and an edge of weight 0 from each column to its immediate successor, plus an edge of weight 0 from the last column of A to the sink. Call the vertex for the first column of A the *source* vertex. Then consider *all* intervals I of at least ℓ columns, test whether the best sequence subset for each I as described above has at least k sequences, and if so, add an edge to G from the first column of I to the immediate successor of the last column of I, weighted by the maximum value of a block with interval I. A *longest path* in the resulting graph G from its source to its sink then gives an optimal packing for A, and the blockiness of A is the length of this longest path divided by the total number of substitutions in A. This graph G would have $\Theta(n^2)$ edges, however, and would not lead to an $O(mn)$ time algorithm for blockiness. Instead, the *only* edges added to G are those for intervals I whose number of columns, or *width*, is in the range $[\ell, 2\ell-1]$. Any block $\mathcal{B}$ whose interval has width at least ℓ is the concatenation of disjoint blocks whose intervals have widths in the above range. Furthermore, the value of block $\mathcal{B}$ is the sum of the values of the blocks in the concatenation. Only adding to G edges in the above width range gives a sparse graph with $O(n)$ vertices and just $O(\ell n)$ edges, which is $O(n)$ edges for constant ℓ.

To implement this algorithm, first construct G in $O(mn)$ time by (1) enumerating the $O(n)$ edges of G in lexicographic order on the pair of column indices defining the column interval for the edge, and then (2) determining the weight of each successive edge e in this order in $O(m)$ time by appending a single column of A to form the column interval for e from the shorter interval of its predecessor. Graph G is acyclic, and a longest source-sink path in a directed acyclic graph can be computed in time linear in its number of vertices and edges [20, pp. 655–657] so the optimal packing in A by blocks can be obtained from G in $O(n)$ time. This takes $O(mn)$ time in total.   □

In practice, there are several minor additions made to this general algorithm to increase its advising accuracy:

(a) Blocks are calculated first the on structure classes α-helix and β-strand. There is an option to then also construct coil blocks on the columns of an alignment that are not already covered by α-helix and β-strand blocks. In practice, including this second coil pass increases the advising accuracy over only including blocks for non-coil classes.

(b) Specify a minimum number of rows k in the definition of a block. Blockiness shows the best performance when this minimum is set to $k = 2$ rows. While using a minimum of $k = 1$ would not have affected the results if we only used $\alpha-$ or $\beta-$blocks, using $k > 1$ increased the number of columns that could be included in coil blocks.

(c) Permitting gap characters in blocks. This allows blocks to be extended to regions that may have single insertions or deletions in one or more sequences. When gaps are allowed in a block they do not contribute to the value of the block (as the value is still defined as the number of residue pairs in the columns and rows of the block), but they can extend a block to include more rows. Including gaps tends to increases the advising accuracy of blockiness.

(d) Define separate α-helix and β-strand block minimum sizes. In actual proteins, α-helixes and β-strands physically both have a minimum number of amino acids to form their structures. There are two modes to capture this: one that sets the minimum based on actual physical sizes and one that sets the minimums to the same length. In the unequal mode, the minimum sizes α-helix, $\ell_\alpha = 4$; β-strand, $\ell_\beta = 3$; and coil, $\ell_c = 2$. In equal mode, $\ell_\alpha = \ell_\beta = \ell_c = 2$. In practice, the unequal mode gives better advising accuracy.

(e) Use the continuous output from PSIPRED. The secondary structure prediction tool PSIPRED outputs confidence values p_t for each structure type $t \in \{\alpha, \beta, c\}$. These can be used to choose a single structure prediction at each position in a protein, by assigning the prediction with the highest confidence value. Alternately, a threshold τ can be used, and thus label a residue in the protein with more than one structure type, but all structure types with $p_t > \tau$. In this way, residues can be in multiple blocks of different structure types if both types have high confidence; in the final packing however, it will only be in one since the blocks of a packing are column-disjoint. In practice, using confidences in this way to allow ambiguous structural types was detrimental to advising accuracy on the benchmarks we considered.

The remaining feature functions in Facet are simpler to compute than Secondary Structure Blockiness.

3.3 Secondary Structure Agreement

As mentioned briefly in the previous section the secondary structure prediction tool PSIPRED [51] outputs confidence values at each residue that are intended to reflect the probability that the residue has each of the three secondary structure types. Denote these three confidences for a residue i, r (the residue in the i-th sequence at the r-th column), normalized so they add up to 1, by $p_\alpha(i, r), p_\beta(i, r)$, and $p_\gamma(i, r)$. Then the probability that two sequences i, j in column r have the same secondary structure type that is not coil can be estimated by looking at the support for that pair from all intermediate sequences k. First define the similarity of two residues (i, r) and (j, r) in column r as

$$S(i, j, r) := p_\alpha(i, r)\, p_\alpha(j, r) + p_\beta(i, r)\, p_\beta(j, r).$$

To measure how strongly the secondary structure locally agrees between sequences i and j around column r, compute a weighted average P of S in a window of width $2\ell + 1$ centered around column r,

$$P(i,j,r) := \sum_{-\ell \leq p \leq \ell} w_p\, S(i,j,r+p)$$

where the weights w_p form a discrete distribution that peaks at $p = 0$ and is symmetric.

Define the support for the pair i, j from intermediate sequence k as the product of the similarities of each i and j with k, $P(i,k,r)\,P(k,j,r)$. The support Q for pair i, j from all intermediate sequences is then defined as

$$Q(i,j,r) := \sum_{\substack{k \\ k \neq i \\ k \neq j}} P(i,k,r)\,P(k,j,r),$$

The value of the Secondary Structure Agreement feature is then the average of $Q(i,j,r)$ over all sequence pairs i, j in all columns r.

This is the feature with the largest running time, but is also one of the strongest features. Its running time is $O(m^3 n\ell)$ for m sequences in an alignment of length n.

The value of ℓ and w must be set by the user. In practice $\ell = 2$ and $w = (0.07, 0.24, 0.38, 0.24, 0.07)$ gave the best advising results.

3.4 Gap Coil Density

A *gap* in a pairwise alignment is a maximal run of either insertions or deletions. For every pair of sequences, there is a set of *gap-residue pairs* (residues that are aligned with gap characters) which each has an associated secondary structure prediction given by PSIPRED (the structure assigned to the residue in the pair). The Gap Coil Density feature measures the fraction of all gap-residue pairs with a secondary structure type of *coil*.

As described, computing Gap Coil Density may seem quadratic in the number of sequences. By simply counting the number of gaps g_i, coil-labeled non-gap entries γ_i, and non-coil-labeled non-gap s_i entries in column i, we can compute this feature by

$$\frac{\displaystyle\sum_{\text{columns } i} g_i\, \gamma_i}{\displaystyle\sum_{\text{columns } i} g_i\, (\gamma_i + s_i)}.$$

All this counting takes linear time total in the number of sequences, so the running time for computing Gap Coil Density is $O(mn)$.

Alternately, PSIPRED confidences can be used; the feature value is then the *average* coil confidences over all gap-residue pairs in the alignment. In practice, using these confidences gives better advising accuracy.

3.5   Gap Extension Density

This feature counts the number of *null characters* in the alignment (the dashes that denote gaps). This is related to affine gap penalties [45], which are commonly used to score alignments. This count is normalized by the total number of alignment entries, or an upper bound U on the number of possible entries. The reason to use an upper bound U is to allow the feature value to be compared across alignments of the same sequences that may have different alignment lengths, while still yielding a feature value that lies in the range $[0, 1]$. This upper bound is calculated as

$$ U := \binom{m}{2} \left(\max_{s \in S} |s| + \max_{s' \in S'} |s'| \right), $$

where $S' := S - \operatorname{argmax}_{s \in S} |s|$. The second part of the sum gives the length of the second longest sequence in S. In practice, normalizing by U gives better advising accuracies.

As the quantity described above is generally *decreasing* in alignment accuracy (since more gaps generally indicates a lower quality alignment), for the actual feature value 1 minus this ratio described above is used.

Gap Extension Density essentially counts the number of *null characters* in an alignment, which can be done in linear time for each sequence. Thus Gap Extension Density can be computed in $O(mn)$ time. The lengths of the input sequences can be computed in linear time, so U can be computed in this same amount of time.

3.6   Gap Open Density

This feature counts the number of *runs* of null characters in the rows of the alignment (which also relates to affine gap penalties). Either the total length of all such runs, or the upper-bound U on alignment size (which tends to give better advising accuracy) can be used for normalization. Just as with Gap Extension Density, Gap Open Density can be computed in $O(mn)$ time.

Similar to Gap Extension Density, the ratio described above is generally decreasing in alignment accuracy, so for the feature value 1 minus the ratio described above is used.

3.7   Gap Compatibility

As in cladistics, we encode the gapping pattern in the columns of an alignment as a binary state: residue (1), or null character (0). For an alignment in this encoding, adjacent columns that have the same gapping pattern can be collapsed into one. This reduced set of columns is evaluated for *compatibility* by checking whether a perfect phylogeny exists on them, using the so-called "four gametes test" on pairs of columns. More specifically, a pair of columns passes the four gametes test if at most three of the four possible patterns 00, 01, 10, 11 occur in the rows of these two columns. A so-called perfect phylogeny exists, in which the binary gapping pattern in each column is explained by a single insertion or deletion event on an edge of the tree, if and only if all pairs of columns pass this test. (See [47, pp. 462–463], or [40].) The Gap Compatibility feature measures the fraction of pairs of columns in the reduced binary data that pass this test, which is a rough measure of how tree-like the gapping pattern is in the alignment. Rather than determining whether a *complete* column pair passes the four-gametes test averaged over all pairs of columns. The *fraction* of a column pair that pass this test (the largest subset of rows that pass the test divided by the total number of rows) can be used as a more fine grained measure. This second version of the feature works better in practice, most likely because it is a less strict measure of the evolutionary compatibility of the gaps.

For each pair of columns, the encoding of each row can be computed in constant time, the counts for the four-gametes states can be collected in linear time in the number of sequences for a given column pair. Since all pairs of columns must be examined, the running time for Gap Compatibility is quadratic in the number of columns. Evaluating this feature takes $O(m^2n)$ time for an alignment with m sequences and n columns.

3.8   Substitution Compatibility

Similar to Gap Compatibility, the substitution pattern in the columns of an alignment can be encoded as binary states: using a reduced amino acid alphabet of equivalency classes, residues in the most prevalent equivalency class in the column are mapped to 1, and all others to 0. This feature measures the fraction of encoded column pairs that pass the four-gametes test, which again is a rough measure of how tree-like the substitution pattern is in the alignment. Just as above there are options for using both whole-column and fractional-column measurements; in practice fractional-column measurements give better accuracy. The standard reduced amino-acid alphabets with 6, 10, 15, and 20 equivalency classes were considered, using the 15-class alphabet gives the strongest correlation with accuracy.

Just like Gap Compatibility, evaluating Substitution Compatibility takes $O(m^2n)$ time.

3.9   Amino Acid Identity

This feature is usually called simply "percent identity." In each induced pairwise alignment, measure the fraction of substitutions in which the residues have the same amino-acid equivalence class. Here the the reduced alphabet with 10 classes is used. The feature averages this fraction over all induced pairwise alignments.

Amino Acid Identity for a whole alignment can be computed by determining the frequency of each amino-acid class in each column, and summing the number of pairs of each alphabet element in a column. Computing amino-acid identity in this way takes $O(mn + n|\Sigma|)$ time for amino-acid equivalence classes Σ. Assuming $|\Sigma|$ is constant, this is $O(mn)$ time.

3.10   Secondary Structure Identity

This feature is like Amino Acid Identity, except instead of the protein's amino-acid sequence, the secondary-structure sequence predicted for the protein by PSIPRED [51], which is a string over the 3-letter secondary structure alphabet, replaces the amino acid equivalence class labels. Similar to the approach described for Amino Acid Identity, Secondary Structure Identity can be computed in $O(mn)$ time.

A variant of this feature uses the secondary structure confidences, where instead of counting identities, we calculate the probability that a pair i, j of residues has the same secondary structure by

$$p_\alpha(i) \, p_\alpha(j) \; + \; p_\beta(i) \, p_\beta(j) \; + \; p_\gamma(i) \, p_\gamma(j).$$

In this version, the prior running-time reduction trick cannot be used, and the algorithm must examine all pairs of rows in a column. This takes total time $O(m^2 n)$ for an alignment with m rows and n columns, but provides better results in practice.

3.11   Average Substitution Score

This feature computes the total score of all substitutions in the alignment, using a BLOSUM62 substitution-scoring matrix [48] which has been shifted and scaled so the amino acid similarity scores are in the range $[0, 1]$. This total score is normalized by the number of substitutions in the alignment, or by upper bound U given earlier so the feature value is comparable between alignments of the same sequences. Normalizing by U provides a feature value that correlates better with true accuracy.

Similar to the running-time reduction of Amino Acid Identity, first count the frequency of each amino acid in each column of an alignment, and sum the BLSM62 score for each possible amino-acid substitution multiplied by the product of the frequency for the two amino acids. This reduces the running time to $O(mn + n|\Sigma|^2)$, which is faster than considering all pairs of rows when $|\Sigma| < m$ (otherwise the naïve $O(m^2n)$ approach can be used).

3.12 Core Column Density

For this feature, *core columns* are predicted as those that only contain residues (and not gap characters) and whose fraction of residues that have the same amino acid equivalency class, for the 10-class alphabet, is above a threshold. The feature then normalizes the count of predicted core columns by the total number of columns in the alignment. The standard reduced alphabets with 6, 10, 15, and 20 equivalency classes were considered, and use the 10-class alphabet is used, as it gave the strongest correlation with true accuracy. Various thresholds were tested and a value of 0.9 was found to give the best trend.

Using the same trick described earlier for Amino Acid Identity, a "core" label can be assigned to the column in linear time, therefore this naïve Core Column Density can be evaluated in $O(mn)$ time. We will later describe a more sophisticated method for predicting core columns in an alignment in Chap. 9.

3.13 Information Content

This feature measures the average entropy of the alignment [49], by summing over the columns the log of the ratio of the abundance of a specific amino acid in the column over the background distribution for that amino acid, normalized by the number of columns in the alignment.

Amino-acid frequencies can be calculated in linear time for each column, and background frequencies for each amino acid can also be found in one pass across the whole alignment. Information content is then evaluated in each column by making one pass over the frequencies for each element in the alphabet. Computing Information Content for an input alignment in $O(mn + m|\Sigma|)$ time for alphabet Σ. Once again, since the alphabet size is constant, this running time is $O(mn)$.

The standard reduced alphabets with 6, 10, 15, and 20 equivalence classes were considered, and the 10-class alphabet is used because it gave the strongest correlation with true accuracy.

3.14   Correlation of Features with Accuracy

Figure 3.1 shows the correlation of each of these features described above with true accuracy. We describe this set of benchmarks and testing procedures in full detail in Chap. 6. Briefly, a total of 861 benchmark alignments was collected from the BENCH suite of [39], which consists of 759 benchmarks, supplemented by a selection of 102 benchmarks from the PALI suite of [9]. For each of these benchmarks the Opal aligner was used to produce a new alignment of the sequences under its default parameter setting. For each of these computed alignments, the underlying

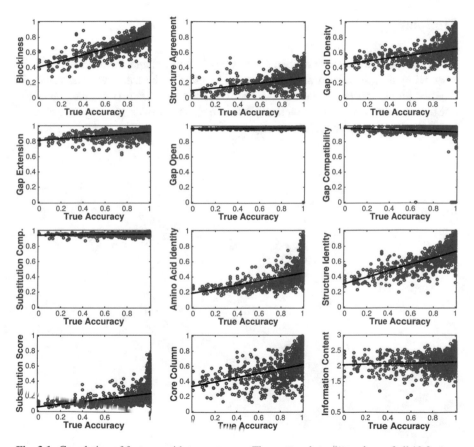

Fig. 3.1 Correlation of features with true accuracy. The scatter plots show values of all 12 feature functions considered for the Facet estimator, on the 861 benchmarks used for testing (described in Chap. 6), using the default parameter setting for the Opal aligner. Each circle is one benchmark alignment plotted with its true accuracy on the horizontal axis (since we know the reference, we can calculate true accuracy) and its feature value on the vertical axis. The line shows is a weighted least-squares line where the weight for a benchmark is calculated to remove the bias towards benchmarks with high accuracy under the default parameter settings. The precise form of the weighting is described in detail in Chap. 6.

correct alignment is known, so the true accuracy of the computed alignment can be obtained. Each of the 12 feature values for each of these alignments was also calculated. The figure shows the correlation of each of the features with true accuracy, where each of the 861 circles in each plot is one benchmark with its true accuracy on the horizontal axis and feature function value on the vertical. Notice that while all of the features generally have a positive trend with true accuracy, the ranges of the feature values differ substantially.

This comprises the set of features considered for constructing the Facet accuracy estimator.

Summary

In this chapter, we have described several easily-computable feature functions for estimating alignment accuracy. Using these features in the framework described in Chap. 2 yields the new accuracy estimator Facet. In Chap. 6 we later give the coefficients for the feature functions, when trained on example alignments from benchmarks with known reference alignments.

4.2   Learning an Optimal Advisor

The optimal advisor problem has several variations, depending on whether the advisor's estimator or set of parameter choices are fixed: (a) simultaneously finding both the best set and estimator, (b) finding the best *set* of parameter choices to use with a given estimator, and (c) finding the best *estimator* to use with a given set of parameter choices. Through out this chapter it is assumed that the features used by the advisor's estimator are given and fixed.

We will first briefly review some concepts that we have discussed in previous chapters and define these terms formally.

From a machine learning perspective, the problem formulations find an advisor that has optimal accuracy on a collection of training data. The underlying training data is

- a suite of *benchmarks*, where each benchmark B_i in the suite consists of a set of sequences to align, together with a *reference alignment* R_i for these sequences that represents their "correct" alignment, and

- a collection of *alternate alignments* of these benchmarks, where each alternate alignment A_{ij} results from aligning the sequences in benchmark i using a parameter choice j that is drawn from a given universe U of parameter choices.

Here a *parameter choice* is an assignment of values to all the parameters of an aligner that may be varied when computing an alignment. Typically an aligner has multiple parameters whose values can be specified, such as the substitution scoring matrix and gap penalties for its alignment scoring function. A parameter choice is represented by a vector whose components assign values to all these parameters. (So for protein sequence alignment, a typical parameter choice is a 3-vector specifying the (1) substitution matrix, (2) gap-open penalty, and (3) gap-extension penalty.) The universe U of parameter choices specifies all the possible parameter choices that might be used for advising. A particular advisor will use a subset $P \subseteq U$ of parameter choices that it considers when advising. In the special case $|P| = 1$, the single parameter choice in set P that is available to the advisor is effectively a *default* parameter choice for the aligner.

Note that since a reference alignment R_i is known for each benchmark B_i, the true accuracy of each alternate alignment A_{ij} for benchmark B_i can be measured by comparing alignment A_{ij} to the reference R_i. Thus for a set $P \subseteq U$ of parameter choices available to an advisor, the most accurate parameter choice $j \in P$ to use on benchmark B_i can be determined in principle by comparing the resulting alternate alignments A_{ij} to R_i and picking the one of highest true accuracy. When aligning sequences in practice, a reference alignment is not known, so an advisor will instead use its estimator to pick the parameter choice $j \in P$ whose resulting alignment A_{ij} has highest *estimated* accuracy.

Chapter 4
Computational Complexity of Advising

This chapter formulates the problem of constructing an optimal advisor: finding both the estimator coefficients and advisor set that give the highest average advising accuracy. Earlier chapters have indicated that both the tasks of independently finding optimal estimator coefficients and optimal advisor sets are NP-complete; the problem of finding both simultaneously is NP-complete as well. This chapter gives NP-completeness proofs for the three problems of Optimal Advisor, Advisor Estimator, and Advisor Set.

4.1 Optimal Parameter Advising

A parameter advisor has two components: (1) the accuracy estimator, which ranks the alternate alignments that the advisor chooses among; and (2) the advisor set, which should be small yet still provide for each input at least one good alternate alignment that the advisor can choose. These two components are very much interdependent. A setting of estimator coefficients may work well for one advisor set, but may not be able to distinguish accurate alignments for another. Similarly, for a given advisor set, one assignment of advisor coefficients may work well, while another may not.

An *optimal advisor* is both an accuracy estimator and an advisor set that together produce the highest average advising accuracy on a collection of benchmarks. In this chapter, we consider the problem of constructing an optimal advisor. We also discuss restrictions of this problem to finding an optimal advisor set for a given estimator, or an optimal estimator for a given advisor set. Finally, we show that all three problems versions are NP-complete.

Adapted from publications [24, 28].

© Springer International Publishing AG 2017   41
D. DeBlasio, J. Kececioglu, *Parameter Advising for Multiple Sequence Alignment*,
Computational Biology 26, https://doi.org/10.1007/978-3-319-64918-4_4

In the problem formulations below, this underlying training data is summarized by

- the *accuracies* a_{ij} of the alternate alignments A_{ij}, where accuracy a_{ij} measures how well the computed alignment A_{ij} agrees with the reference alignment R_i, and

- the *feature vectors* F_{ij} of these alignments A_{ij}, where each vector F_{ij} lists the values for A_{ij} of the estimator's feature functions.

As defined in Chap. 2, for an estimator that uses t feature functions, each feature vector F_{ij} is a vector of t feature values,

$$F_{ij} = (g_{ij1}\ g_{ij2}\ \cdots\ g_{ijt}),$$

where each feature value g_{ijh} is a real number satisfying $0 \leq g_{ijh} \leq 1$. Feature vector F_{ij} is used by the advisor to evaluate its accuracy estimator E on alignment A_{ij}. Let the *coefficients* of the estimator E be given by vector

$$c = (c_1\ c_2\ \cdots\ c_t).$$

Then the value of accuracy estimator E on alignment A_{ij} is given by the inner product

$$E_c(A_{ij}) = c \cdot F_{ij} = \sum_{1 \leq h \leq t} c_h\, g_{ijh}. \tag{4.1}$$

Informally, the objective function that the problem formulations seek to maximize is the average accuracy achieved by the advisor across the suite of benchmarks in the training set. The benchmarks may be nonuniformly weighted in this average to correct for bias in the training data, such as the over-representation of easy benchmarks that typically occurs in standard benchmark suites.

A subtle issue that the formulations must take into account is that when an advisor is selecting a parameter choice via its estimator, there can be ties in the estimator value, so there may not be a unique parameter choice that maximizes the estimator. In this situation, it can be assumed that the advisor *randomly* selects a parameter choice among those of maximum estimator value. Given this randomness, the performance of an advisor on an input is measured by its *expected* accuracy on that input.

Furthermore, in practice any accuracy estimator inherently has *error* (otherwise it would be equivalent to true accuracy), and a robust formulation for learning an advisor should be tolerant of error in the estimator. Let $\epsilon \geq 0$ be a given error *tolerance*, and P be the set of parameter choices used by an advisor. We define the set $\mathcal{O}_i(P)$ of parameter choices that the advisor could potentially *output* for benchmark B_i as

$$\mathcal{O}_i(P) = \left\{ j \in P : E_c(A_{ij}) \geq e_i^* - \epsilon \right\}, \tag{4.2}$$

where $e_i^* := \max\{E_c(A_{i\tilde{j}}) \ : \ \tilde{j} \in P\}$ is the maximum estimator value on benchmark B_i. The parameter choice output by an advisor on benchmark B_i is selected uniformly at random among those in $\mathcal{O}_i(P)$. Note that when $\epsilon = 0$, set $\mathcal{O}_i(P)$ is simply the set of parameter choices that are tied for maximizing the estimator. A nonzero tolerance $\epsilon > 0$ can aid in learning an advisor that has improved generalization to testing data.

The *expected accuracy* achieved by the advisor on benchmark B_i using set P is then

$$\mathcal{A}_i(P) \ = \ \frac{1}{|\mathcal{O}_i(P)|} \sum_{j \in \mathcal{O}_i(P)} a_{ij}. \tag{4.3}$$

In learning an advisor, the objective is to find a set P that maximizes the advisor's expected accuracy $\mathcal{A}_i(P)$ on the training benchmarks B_i.

Formally, an optimal advisor is one that maximizes the following *objective function*,

$$f_c(P) \ = \ \sum_i w_i \, \mathcal{A}_i(P), \tag{4.4}$$

where i indexes the benchmarks, and w_i is the weight placed on benchmark B_i. (The benchmark weights are to correct for possible sampling bias in the training data.) In words, objective $f_c(P)$ is the expected accuracy of the parameter choices selected by the advisor averaged across the weighted training benchmarks, using advisor set P and the estimator given by coefficients c. The objective function is written as as $f(P)$ without subscript c when the estimator coefficient vector c is fixed or understood from context.

The following *argmin* and *argmax* notation are used later. For a function f and a subset S of its domain,

$$\mathrm{argmin}\{f(x) \ : \ x \in S\}$$

denotes the set of *all* elements of S that achieve the minimum value of f, or in other words, the set of minimizers of f on S. Similarly, argmax is used to denote the set of maximizers.

4.2.1 Optimal Advisor

The problem of finding an *optimal advisor* involves simultaneously finding an advisor estimator and an advisor set that together yields the highest average advising accuracy.

In the problem definition,

- n is the number of benchmarks, and
- t is the number of alignment features.

Set $\mathcal{Q}$ denotes the set of rational numbers.

Definition 4.1 (Optimal Advisor) The *Optimal Advisor* problem takes as input

- cardinality bound $k \geq 1$,
- universe U of parameter choices,
- weights $w_i \in \mathcal{Q}$ on the training benchmarks B_i, where each $w_i \geq 0$ and $\sum_i w_i = 1$,
- accuracies $a_{ij} \in \mathcal{Q}$ of the alternate alignments A_{ij}, where each $0 \leq a_{ij} \leq 1$,
- feature vectors $F_{ij} \in \mathcal{Q}^t$ for the alternate alignments A_{ij}, where each feature value g_{ijh} in vector F_{ij} satisfies $0 \leq g_{ijh} \leq 1$, and
- error tolerance $\epsilon \in \mathcal{Q}$ where $\epsilon \geq 0$.

The output is

- estimator coefficient vector $c \in \mathcal{Q}^t$, where each coefficient c_i in vector c satisfies $c_i \geq 0$ and $\sum_{1 \leq i \leq t} c_i = 1$, and
- set $P \subseteq U$ of parameter choices for the advisor, with $|P| \leq k$,

that maximizes objective $f_c(P)$ given by Eq. (4.4). $\qquad\qquad\qquad\qquad$ $\square$

4.2.2 Advisor Set

The optimal advisor problem can be restricted to finding an *optimal set* of parameter choices for advising with a given estimator.

Definition 4.2 (Advisor Set) The *Advisor Set* problem takes as input

- weights w_i on the benchmarks,
- accuracies a_{ij} of the alternate alignments,
- feature vectors F_{ij} for the alternate alignments,
- coefficients $c = (c_1 \cdots c_t) \in \mathcal{Q}^t$ for the estimator, where each $c_i \geq 0$ and $\sum_{1 \leq i \leq t} c_i = 1$, and
- error tolerance ϵ.

The output is

- advisor set P

that maximizes objective $f_c(P)$ given by Eq. (4.4). $\qquad\qquad\qquad\qquad$ $\square$

4.2.3 Advisor Estimator

Similarly, the problem of finding an *optimal estimator* can be defined where the set of parameter choices for the advisor is now given.

Definition 4.3 (Advisor Estimator) The *Advisor Estimator* problem takes as input

- weights w_i on the benchmarks,
- accuracies a_{ij} of the alternate alignments,
- feature vectors F_{ij} for the alternate alignments,
- advisor set P, and
- error tolerance ϵ.

The output is

- coefficients $c = (c_1 \cdots c_t) \in \mathcal{Q}^t$ for the estimator, where each $c_i \geq 0$ and $\sum_{1 \leq i \leq t} c_i = 1$,

that maximize objective $f_c(P)$ given by Eq. (4.4).   □

For Advisor Estimator, resolving ties to pick the worst among the parameter choices that maximize the estimator, as in the definition of $\mathcal{A}(i)$ in Eq. (4.4), is crucial, as otherwise the problem formulation becomes degenerate. If the advisor is free to pick any of the tied parameter choices, it can pick the tied one with highest true accuracy; if this is allowed, the optimal estimator c^* that is found by the formulation would degenerate to the flattest possible estimator that evaluates all parameter choices as equally good (since the degenerate flat estimator would make the advisor appear to match the performance of a perfect oracle on set P). Resolving ties in the worst-case way eliminates this degeneracy.

4.3   Complexity of Learning Optimal Advisors

In this section we prove that Advisor Set, the problem of learning an optimal parameter set for an advisor (given by Definition 4.2 of Sect. 4.2) is NP-complete, and hence unlikely to be efficiently solvable in the worst-case. As is standard, NP-completeness is proved for a decision version of this optimization problem, which is a version whose output is a yes/no answer (as opposed to a solution that optimizes an objective function).

The *decision version* of Advisor Set has an additional input $\ell \in \mathcal{Q}$, which will lower bound the objective function. The decision problem is to determine, for the input instance $k, U, w_i, a_{ij}, F_{ij}, c, \epsilon, \ell$, whether or not there exists a set $P \subseteq U$ with $|P| \leq k$ for which the objective function has value $f_c(P) \geq \ell$.

Theorem 4.1 (NP-Completeness of Advisor Set) *The decision version of Advisor Set is NP-complete.*

Proof. A reduction can be made from the *Dominating Set* problem, which is NP-complete [44, problem GT2], as follows. The input to Dominating Set is an undirected graph $G = (V, E)$ and an integer k, and the problem is to decide whether or not G contains a vertex subset $S \subseteq V$ with $|S| \leq k$ such that every vertex in V is in S or is adjacent to a vertex in S. Such a set S is called a *dominating set* for G.

Given an instance G, k of Dominating Set, construct an instance $U, w_i, a_{ij}, F_{ij}, c,$ ϵ, ℓ of the decision version of Advisor Set as follows. For the cardinality bound use the same value k, for the number of benchmarks take $n = |V|$, and index the universe of parameter choices by $U = \{1, \dots, n\}$; have only one feature $(d = 1)$ with estimator coefficients $c = 1$; use weights $w_i = 1/n$, error tolerance $\epsilon = 0$, and lower bound $\ell = 1$. Let the vertices of G be indexed $V = \{1, \dots, n\}$. (So both the set of benchmarks and the universe of parameter choices in essence correspond to the set of vertices V of graph G.) Define the *neighborhood* of vertex i in G to be $N(i) := \{j : (i,j) \in E\} \cup \{i\}$, which is the set of vertices adjacent to i, including i itself. For the alternate alignment accuracies, take $a_{ij} = 1$ when $j \in N(i)$; otherwise, $a_{ij} = 0$. For the feature vectors, assign $F_{ij} = a_{ij}$.

The claim is then that G, k is a yes-instance of Dominating Set iff $k, U, w_i, a_{ij}, F_{ij}, c, \epsilon, \ell$ is a yes-instance of Advisor Set.

To show the forward implication, suppose G has a dominating set $S \subseteq V$ with $|S| \leq k$, and consider the advisor set $P = S$. With the above construction, for every benchmark, set $\mathcal{O}_i(P) = N(i) \cap S$, which is nonempty (since S is a dominating set for G). So $\mathcal{A}_i(P) = 1$ for all benchmarks. Thus for this advisor set P, the objective function has value $f_c(P) = 1 \geq \ell$.

For the reverse implication, suppose advisor set P achieves objective value $\ell = 1$. Since P achieves value 1, for every benchmark it must be that $\mathcal{A}_i(P) = 1$. By construction of the a_{ij}, this implies that in G every vertex $i \in V$ is in P or is adjacent to a vertex in P. Thus set $S = P$, which satisfies $|S| \leq k$, is a dominating set for G.

This reduction shows Advisor Set is NP-hard, as the instance of Advisor Set can be constructed in polynomial time. Furthermore, it is in NP, as we can nondeterministically guess an advisor set P, and then check whether its cardinality is at most k and its objective value is at least ℓ in polynomial time. Thus Advisor Set is NP-complete.   □

Note that the proof of Theorem 4.1 shows Advisor Set is NP-complete for the special case of a *single feature*, error tolerance zero, when all accuracies and feature values are binary, and benchmarks are uniformly weighted.

In general, the goal is to to find an optimal *parameter advisor*, which requires simultaneously finding both the best possible parameter *set* and the best possible accuracy *estimator*. The general problem of constructing an optimal parameter advisor is defined as follows.

The *decision version* of Optimal Advisor, similar to the decision version of Advisor Set, has an additional input ℓ that lower bounds the objective function.

We next prove that Optimal Advisor is NP-complete. While its NP-hardness follows from Advisor Set, the difficulty is in proving that this more general problem is still in the class NP.

Theorem 4.2 (NP-Completeness of Optimal Advisor) *The decision version of Optimal Advisor is NP-complete.*

Proof. The proof of Theorem 4.1 shows Advisor Set remains NP-hard for the special case of a single feature. To prove the decision version of Optimal Advisor is NP-hard, use *restriction*: simply reduce Advisor Set with a single feature to Optimal

Advisor (reusing the instance of Advisor Set for Optimal Advisor). On this restricted input with $d = 1$, Optimal Advisor is equivalent to Advisor Set, so Optimal Advisor is also NP-hard.

The two proofs above can be combined to show that the general Optimal Advisor problem is in class NP. To decide whether its input is a yes-instance, after first nondeterministically guessing parameter set $P \subseteq U$ with $|P| \leq k$, make for each benchmark i a nondeterministic guess for its sets $\mathcal{O}_i(P)$ and $\mathcal{M}_i(P) := \operatorname{argmax} \{c \cdot F_{ij} : j \in P\}$, *without* yet knowing the coefficient vector c. Call $\widetilde{O}_i$ the guess for set $\mathcal{O}_i(P)$, and $\widetilde{M}_i$ the guess for set $\mathcal{M}_i(P)$, where $\widetilde{M}_i \subseteq \widetilde{O}_i \subseteq P$. To check whether a coefficient vector c exists that satisfies $\mathcal{O}_i(P) = \widetilde{O}_i$ and $\mathcal{M}_i(P) = \widetilde{M}_i$, then construct the following linear program with variables $c = (c_1 \cdots c_d)$ and ξ. The objective function for the linear program is to maximize the value of variable ξ. The constraints are: $c_h \geq 0$ and $\sum_{1 \leq h \leq d} c_h = 1$; $0 \leq \xi \leq 1$; for all benchmarks i and all parameter choices $j^* \in \widetilde{M}_i$ and $j \notin \widetilde{M}_i$,

$$c \cdot F_{ij^*} \geq c \cdot F_{ij} + \xi ;$$

for all benchmarks i and all parameter choices $j, \tilde{j} \in \widetilde{M}_i$,

$$c \cdot F_{ij} = c \cdot F_{i\tilde{j}} ;$$

for all benchmarks and all parameter choices $j^* \in \widetilde{M}_i$ and $j \in \widetilde{O}_i$,

$$c \cdot F_{ij} \geq c \cdot F_{ij^*} - \epsilon .$$

This linear program can be solved in polynomial time. If it has a feasible solution, then it has an optimal solution (as its objective function is bounded). In an optimal solution c^*, ξ^* we check whether $\xi^* > 0$. If this condition holds, the guessed sets $\widetilde{O}_i, \widetilde{M}_i$, correspond to actual sets $\mathcal{O}_i(P)$ and $\mathcal{M}_i(P)$ for an estimator. For each benchmark i, we then evaluate $\mathcal{A}_i(P)$, and check whether $\sum_i w_i \mathcal{A}_i(P) \geq \ell$. Note that after guessing the sets P, $\widetilde{O}_i$, and $\widetilde{M}_i$, the rest of the computation runs in polynomial time. Thus Optimal Advisor is in NP. $\square$

Finally we show the NP-completeness of the decision version of Advisor Estimator.

Theorem 4.3 (NP-Completeness of Advisor Estimator) *The decision version of Advisor Estimator is NP-complete.*

Proof. To show Advisor Estimator is NP-hard, use a similar reduction from Dominating Set that was used in proving Theorem 4.1. Given an instance G, k of Dominating Set, first construct an instance $w_i, a_{ij}, F_{ij}, P, \epsilon, \ell, \delta$ of the decision version of Advisor Estimator, using the same cardinality bound k, number of benchmarks and parameter choices $n = |V|$, weights $w_i = 1/n$, error tolerance $\epsilon = 0$, accuracies a_{ij} again defined as before, and lower bound $\ell = 1$, as for Advisor Set. For the set P of parameter choices for the advisor, take $P = \{1, \ldots, n\}$. The number of features is now $t = n$. (So in essence the set of benchmarks, the advisor set, and the set of features all coincide with the set of vertices V.) For the feature vectors take $F_{ij} = (0 \cdots 0\, a_{ij}\, 0 \cdots 0)$ which has value a_{ij} at location j. This is equivalent to a feature vector F_{ij} that is all zeroes, except for a 1 at location j if $j = i$ or vertex j is adjacent to vertex i in G. For the precision lower bound we take $\delta = 1/k$. Note that this instance of Advisor Estimator can be constructed in polynomial time.

We claim G, k is a yes-instance of Dominating Set iff $w_i, a_{ij}, F_{ij}, P, \epsilon, \ell, \delta$ is a yes-instance of Advisor Estimator. To show the reverse implication, first notice that with the chosen δ, coefficient vector c can have at most k nonzero coefficients (since if c has more than k nonzero coefficients, $\sum_i c_i > k\delta = 1$, a contradiction). Let feature subset $S \subseteq V$ be all indices i at which $c_i > 0$. Call S the *support* of c, and by our prior observation $|S| \leq k$. By construction of the feature vectors, $c \cdot F_{ij} = c_j$ if $j \in N(i)$; otherwise, $c \cdot F_{ij} = 0$. This further implies that $\mathcal{A}_i(P) = 1$ if $S \cap N(i)$ is nonempty; otherwise, $\mathcal{A}_i(P) = 0$. So if there exists coefficient vector c such that the objective function achieves value 1, then the support S of c gives a vertex subset $S \subseteq V$ that is a dominating set for G. For the forward implication, given a dominating set $S \subseteq V$ for G, take for the estimator coefficients $c_i = 1/|S|$ if $i \in S$, and $c_i = 0$ otherwise. The nonzero coefficients of this vector c have value at least δ, and by the same reasoning as above, each $\mathcal{A}_i(P) = 1$ as S is a dominating set, so the estimator given by this vector c yields an advisor that achieves objective value 1, which proves the claim.

To show Advisor Estimator is in class NP, use the same construction used for proving Optimal Advisor is in class NP. For each benchmark we can make a nondeterministic choice for its set $\widetilde{\mathcal{O}}_i(P)$, and compute $\widetilde{\mathcal{M}}_i$. Then construct a linear program to determine if these guesses are actual sets for the estimator. The guesses and solution of the linear program can be performed in polynomial time. Thus Advisor Estimator is in NP.   □

Summary

This chapter formulated the problem of finding an optimal advisor, and two related problems of finding an optimal advisor set and an optimal accuracy estimator. We then proved that all three problems are NP-complete. The next chapter describes practical approaches to the Advisor Set problem, and how to model these three problems by mixed-integer linear programming.

Chapter 5
Constructing Advisors

This chapter considers how to find sets of parameter choices for an advisor. We consider two forms of advisor sets: (1) estimator-oblivious sets that are optimal for a perfect advisor called an oracle, and (2) estimator-aware sets that are optimal for a given accuracy estimator. Chapter 4 proved that learning an optimal advisor set is NP-complete. Here we model finding optimal advisor sets as an integer linear program. In practice this integer program cannot be solved to optimality, so we develop an efficient approximation algorithm that finds near-optimal sets, and prove a tight bound on its approximation ratio.

Looking Ahead

Chapter 4 formulated the *optimal advisor* problem and proved that it is NP-complete. In this chapter, we show how to model finding optimal advisors by integer linear programming. In practice, though, these integer linear programming models are not solvable to optimality, even for very small inputs.

Since Sect. 2.2.2 gave a practical method for finding estimator coefficients for a set of examples, we might concentrate instead on the restricted *advisor set* problem. As shown previously though, this problem too is NP-complete, and the restricted integer program is also not solvable in practice.

On the other hand, a special version of the integer program for advisor sets where the advisor knows the true accuracy of alignments is solvable to optimality. We call such an advisor set for this perfect advisor an *oracle set*. While in practice we do not actually know true accuracy, an optimal oracle set can be used with an accuracy estimator to yield a decent advisor.

Adapted from publications [24, 26].

© Springer International Publishing AG 2017
D. DeBlasio, J. Kececioglu, *Parameter Advising for Multiple Sequence Alignment*,
Computational Biology 26, https://doi.org/10.1007/978-3-319-64918-4_5

The last section of this chapter then focuses on how to find, in a reasonable amount of time, good advisor sets that are tailored to the actual accuracy estimator used by a realistic advisor. In particular, we develop an efficient greedy *approximation algorithm* that is guaranteed to find near-optimal advisor sets for a given estimator. Experiments in later chapters show that the *greedy* advisor sets found by this approximation algorithm, using Facet, TCS, MOS, PredSP, or GUIDANCE as the accuracy estimator, outperform optimal *oracle* sets at all cardinalities. Furthermore, on training data, for some estimators these suboptimal greedy sets perform surprisingly close to optimal *exact* sets found by exhaustive search, and moreover, these greedy sets actually generalize better than exact sets. As a consequence, on testing data, for some estimators the greedy sets output by the approximation algorithm actually give superior performance to exact sets for parameter advising.

5.1 Integer Linear Programs for Optimal Advising

Recall that an *integer linear program* (ILP) is an optimization problem with a collection of integer-valued variables, an objective function to optimize that is linear in these variables, and constraints that are linear inequalities in the variables. The formulations of Advisor Coefficients and Optimal Advisor are actually so-called *mixed-integer* programs, where some of the variables are real-valued, while Advisor Set has all integer variables.

The integer linear programming formulations given below actually model a more general version of the advising problems. The advising problems in Sect. 4.2 define the advisor $\mathcal{A}$ so that it carefully resolves ties among the parameter choices that achieve the optimum value of the estimator, by picking from this tied set the parameter choice that has lowest true accuracy. (This finds a solution that has the best possible average accuracy, even in the worst case.) This definition of advisor $\mathcal{A}$ is extended here to now pick from a larger set of *near-optimal* parameter choices with respect to the estimator. To make this precise, for benchmark i, set P of parameter choices, and a real-value $\delta \geq 0$, let

$$M_\delta(i) \ := \ \left\{ j \in P \ : \ c \cdot F_{ij} \geq \max_{k \in P} \{c \cdot F_{ik}\} - \delta \right\}.$$

Set $M_\delta(i)$ is the near-optimal parameter choices that are within δ of maximizing the estimator for benchmark i. (So $M_\delta(i) \supseteq \mathrm{argmax}_{j \in P}\{c \cdot F_{ij}\}$, with equality when $\delta = 0$.) The definition of the advisor $\mathcal{A}$ is then extended in Eq. (4.4) for $\delta \geq 0$ to

$$\mathcal{A}(i) \ \in \ \mathrm{argmin}\left\{a_{ij} \ : \ j \in M_\delta(i)\right\}. \tag{5.1}$$

At $\delta = 0$, this coincides with the original problem definitions. The extension to $\delta > 0$ is designed to boost the *generalization* of optimal solutions (in other words, to find a solution that is not overfitted to the training data), when cross-validation experiments are performed on independent training and test sets as in Chap. 6.

5.1.1 Modeling Advisor Set

The integer linear program (ILP) for Advisor Set has three classes of *variables*, which all take on binary values $\{0, 1\}$. Variables x_{ij}, for all benchmarks i and all parameter choices j from the universe, encode the advisor $\mathcal{A}$: $x_{ij} = 1$ if the advisor uses choice j on benchmark i; otherwise, $x_{ij} = 0$. Variables y_j, for all parameter choices j from the universe, encode the set P that is found by Advisor Set: $y_j = 1$ iff $j \in P$. Variables z_{ij}, for all benchmarks i and parameter choices j, encode the parameter choice in P with highest estimator value for benchmark i: if $z_{ij} = 1$ then $j \in \mathrm{argmax}_{k \in P} \ c \cdot F_{ik}$. This argmax set may contain several choices j, and in this situation the ILP given below arbitrarily selects one such choice j for which $z_{ij} = 1$.

For convenience, the description of the ILP below also refers to the new *constants* e_{ij}, which are the estimator values of the alternate alignments A_{ij}: for the fixed estimator c for Advisor Set, $e_{ij} = c \cdot F_{ij}$.

The *objective function* for the ILP is to maximize

$$\sum_i w_i \sum_j a_{ij} x_{ij}. \tag{5.2}$$

In this function, the inner sum $\sum_j a_{ij} x_{ij}$ will be equal to $a_{i,\mathcal{A}(i)}$, as the x_{ij} will capture the (unique) parameter choice that advisor $\mathcal{A}$ makes for benchmark i. This objective is linear in the variables x_{ij}.

The *constraints* for the ILP fall into three classes. The *first class* ensures that variables y_j encode set P, and variables x_{ij} encode an assignment to benchmarks from P. The ILP has constraints

$$\sum_j y_j \ \leq \ k, \tag{5.3}$$

$$\sum_j x_{ij} \ = \ 1, \tag{5.4}$$

$$x_{ij} \ \leq \ y_j, \tag{5.5}$$

where Eq. (5.4) occurs for all benchmarks i, and inequality (5.5) occurs for all benchmarks i and all parameter choices j.

In the above, inequality (5.3) enforces $|P| \leq k$. Equation (5.4) force the advisor to select one parameter choice for every benchmark. Inequalities (5.5) enforce that the advisor's selections must be parameter choices that are available in P.

The *second class* of constraints ensure that variables z_{ij} encode a parameter choice from P with highest estimator value. To enforce that the z_{ij} encode an assignment to benchmarks from P,

$$\sum_j z_{ij} = 1, \tag{5.6}$$

$$z_{ij} \leq y_j, \tag{5.7}$$

where Eq. (5.6) occurs for all i, and inequality (5.7) occurs for all i and j. (In general, the z_{ij} will differ from the x_{ij}, as the advisor does not necessarily select the parameter choice with highest estimator value.) For all benchmarks i, and all parameter choices j and k from the universe with $e_{ik} < e_{ij}$, we have the inequality

$$z_{ik} + y_j \leq 1. \tag{5.8}$$

Inequalities (5.8) ensure that if a parameter choice k is identified as having the highest estimator value for benchmark i by $z_{ik} = 1$, there must not be any other parameter choice j in P that has higher estimator value on i. Note that the constants e_{ij} are known in advance, so inequalities (5.8) can be enumerated by sorting all j by their estimator value e_{ij}, and collecting the ordered pairs (k, j) from this sorted list.

The *third class* of constraints ensure that the parameter choices x_{ij} selected by the advisor correspond to the definition in Eq. (5.1): namely, among the parameter choices in P that are within δ of the highest estimator value from P for benchmark i, the parameter choice of lowest accuracy is selected. For all benchmarks i, all parameter choices j, and all parameters choices k and h with both $e_{ik}, e_{ih} \in [e_{ij}-\delta, e_{ij}]$ and $a_{ih} < a_{ik}$, we have the inequality

$$x_{ik} + y_h + z_{ij} \leq 2. \tag{5.9}$$

Inequalities (5.9) ensure that for the parameter choices that are within δ of the highest estimator value for benchmark i, the advisor only selects parameter choice k for i if k is within δ of the highest and there is no other parameter choice available in P within δ of the highest that has lower accuracy. Finally, for all benchmarks i and all parameter choices j and k with $e_{ik} < e_{ij}-\delta$, we have the inequality

$$x_{ik} + y_j \leq 1. \tag{5.10}$$

Inequalities (5.10) enforce that the advisor cannot select parameter choice k for i if the estimator value for k is below δ of an available parameter choice in P. (Inequalities (5.9) capture the requirements on parameter choices that are within δ of the highest, while inequalities (5.10) capture the requirements on parameter choices that are below δ of the highest.)

A truly remarkable aspect of this formulation is that the ILP is able to capture all the subtle conditions the advisor must satisfy through its static set of inequalities (listed at "compile time"), *without* knowing when the ILP is written what the optimal *set P* is, and hence without knowing what parameter choices in *P* have the highest estimator value for each benchmark.

To summarize, the ILP for Advisor Set has binary variables x_{ij}, y_j, and z_{ij}, and inequalities (5.3) through (5.10). For n benchmarks and a universe of m parameter choices, this is $O(mn)$ variables, and $O(m^2n + m\widetilde{m}^2n)$ constraints, where $\widetilde{m}$ is the maximum number of parameter choices that are within δ in estimator value of any given parameter choice. For small δ, typically $\widetilde{m} \ll m$, which leads to $O(m^2n)$ constraints in practice. In the worst-case, though, the ILP has $\Theta(m^3n)$ constraints.

An alternate ILP formulation can be constructed that adds $O(n)$ real-valued variables to capture the highest estimator value from P for each benchmark i, and only has $O(m^2n)$ total constraints (so fewer constraints than the above ILP in the worst case), but its objective function is more involved, and attempts to solve the alternate ILP suffered from numerical issues.

5.1.2 Modeling Advisor Estimator

The following modifications of the above ILP for Advisor Set are needed to obtain an ILP for Advisor Estimator: (a) the set P is now fixed; and (b) the estimator c is no longer fixed, so the enumeration of inequalities cannot exploit concrete estimator values. The fact that that set P is now part of the input can be handled easily by appropriately fixing the variables y_j with new equations: for all $j \in P$ add equation $y_j = 1$, and for all $j \notin P$ add $y_j = 0$.

To find the optimal estimator c, add ℓ new real-valued variables $c_1, \ldots, c_\ell$ with the constraints $\sum_h c_h = 1$ and $c_h \geq 0$. Two new classes of binary-valued integer variables also need to be added: (a) variable s_{ij}, for all benchmarks i and all parameter choices j, which has value 1 when the estimator value of parameter choice j on benchmark i, namely $c \cdot F_{ij}$, is within δ of the highest estimator value for i; and (b) variable t_{ijk}, for all benchmarks i and all parameter choices j and k, which has value 1 when $c \cdot F_{ij} > c \cdot F_{ik} - \delta$.

To set the values of the binary variables t_{ijk}, for all i, j, k we add the inequalities

$$t_{ijk} \geq c \cdot F_{ij} - c \cdot F_{ik} + \delta. \tag{5.11}$$

This inequality is linear in the variables $c_1, \ldots, c_\ell$. Note that the value of the estimator $c \cdot F_{ij}$ will always be in the range $[0, 1]$, and we are assuming that the constant $\delta \ll 1$. To set the values of the binary variables s_{ij}, for all i, j, k we add the inequalities

$$s_{ij} \geq t_{ijk} + t_{ikj} + z_{ik} - 2. \tag{5.12}$$

While the ILP only has to capture relationships between parameter choices that are in set P, we do not constrain the variables s_{ij} and t_{ijk} for parameter choices outside P to be 0, but allow the ILP to set them to 1 if needed for a feasible solution.

The variables s_{ij} and t_{ijk} are now used to express the relationships in the former inequalities (5.8) through (5.10). Inequality (5.8) is replaced by the following inequality over all i, j, k,

$$z_{ij} + y_k \leq 2 - \left(c \cdot F_{ik} - c \cdot F_{ij} \right). \tag{5.13}$$

Inequality (5.9) is replaced by the following inequality over all i, j, k with $a_{ik} < a_{ij}$,

$$x_{ij} + y_k + s_{ij} + s_{ik} \leq 3. \tag{5.14}$$

Finally, we replace inequality (5.10) is replaced by the following inequality over all i, j, k,

$$x_{ij} + y_k \leq 2 - \left(c \cdot F_{ik} - c \cdot F_{ij} - \delta \right). \tag{5.15}$$

To summarize, the ILP for Advisor Estimator has binary variables x_{ij}, y_j, z_{ij}, s_{ij}, t_{ijk}, real variables c_h, constraints (5.6)–(5.7) and (5.11)–(5.15), plus the elementary constraints on the y_j and c_h. This is $O(m^2 n)$ variables and $O(m^2 n)$ constraints. While in general this is an enormous mixed-integer linear program, it can be solved to optimality for small, fixed sets P. Its difficulty increases with the size of P, and instances up to $|P| \leq 4$ can be solved in two days of computation.

5.1.3   Modeling Optimal Advisor

The ILP for Optimal Advisor is simply the above ILP for Advisor Estimator where set P coincides with the entire universe of parameter choices: $P = \{1, \ldots, m\}$. Solving this ILP is currently beyond reach.

While very large integer linear programs can be solved to optimality in practice using modern solvers such as CPLEX [50], there is no known algorithm for integer linear programming that is efficient in the worst-case. Thus our reductions of the optimal advising problems to integer linear programming do not yield algorithms for these problems that are guaranteed to be efficient. On the other hand, as Sect. 4.3 showed, the optimal advising problems are all NP-complete, so it is unlikely that *any* worst-case efficient algorithm for them exists.

5.2   Constructing Optimal Oracle Sets

While the goal is to find advisor sets that are optimal for the actual accuracy estimator used by an advisor, in practice finding such optimal set is intractable for all but the smallest cardinalities. Nevertheless, it is possible in practice to find optimal advisor sets that are *estimator oblivious*, in the sense that the set-finding algorithm is unaware of the mistakes made by the advisor due to using an accuracy estimator rather than knowing true accuracy. More precisely, we can find an optimal advisor set can be found for an advisor whose "estimator" is the true accuracy of an alignment. As mentioned previously, we call such a perfect advisor an *oracle*.

To find an optimal oracle set, use the same objective function described in Eq. (5.2), and Eqs. (5.3) through (5.5) to make sure an alignment is only selected if the parameter that is used to generate it is chosen. Solving the ILP with only these constraints will yield an optimal advisor set for the oracle advisor. Note that ties in accuracy do not need to be resolved, as any alignment with a tied "estimator" value also has a tied accuracy value, and thus would not effect the objective value.

With the reduced number of variables and constraints in this modified ILP, optimal oracle sets can be found in practice even for very large set cardinalities.

5.3 Constructing Near-Optimal Advisor Sets

As Advisor Set is NP-complete, it is unlikely we can efficiently find advisor sets that are *optimal*; however, advisor sets that are guaranteed to be *close* to optimal can be found efficiently, in the following sense. An α-*approximation algorithm* for a maximization problem, where $\alpha < 1$, is a polynomial-time algorithm that finds a feasible solution whose value under the objective function is at least factor α times the value of an optimal solution. Factor α is called the *approximation ratio*. In this section, we show that for any constant ℓ with $\ell \leq k$, there is a simple approximation algorithm for Advisor Set that achieves approximation ratio ℓ/k.

For constant ℓ, the optimal advisor set of cardinality at most ℓ can be found in polynomial time by exhaustive search (since when ℓ is a constant there are polynomially-many subsets of size at most ℓ). The following natural approach to Advisor Set builds on this idea, by starting with an optimal advisor set of size at most ℓ, and greedily augmenting it to one of size at most k. Since augmenting an advisor set by adding a parameter choice can worsen its value under the objective function, even if augmented in the best possible way, the procedure Greedy given below outputs the best advisor set found across all cardinalities.

procedure Greedy(ℓ, k) **begin**
    Find an optimal subset $P \subseteq U$ of size $|P| \leq \ell$ that maximizes $f(P)$.
    $(\widetilde{P}, \widetilde{\ell}) := (P, |P|)$
    for cardinalities $\widetilde{\ell}+1, \ldots, k$ **do begin**
        Find parameter choice $j^* \in U - \widetilde{P}$ that maximizes $f(\widetilde{P} \cup \{j^*\})$.
        $\widetilde{P} := \widetilde{P} \cup \{j^*\}$
        if $f(\widetilde{P}) > f(P)$ **then** $P := \widetilde{P}$
    end
    output P
end

This natural greedy procedure is an approximation algorithm for Advisor Set as shown below.

Theorem 5.1 (Approximation Ratio) *Procedure* Greedy *is an* (ℓ/k)-*approximation algorithm for Advisor Set with cardinality bound k, and any constant ℓ with $\ell \leq k$.*

Proof. The basic idea of the proof is to use averaging over all subsets of size ℓ from the optimal advisor set of size at most k, in order to relate the objective function value of the set found by Greedy to the optimal solution.

To prove the approximation ratio, let

- P^* be the optimal advisor set of size at most k,
- $\widetilde{P}$ be the optimal advisor set of size at most ℓ,
- P be the advisor set output by Greedy,
- $\mathcal{S}$ be the set of all subsets of P^* that have size ℓ,
- $\widetilde{k}$ be the size of P^*, and
- $\widetilde{\ell}$ be the size of $\widetilde{P}$.

Note that if $\widetilde{k} < \ell$, then the greedy advisor set P is actually optimal and the approximation ratio holds. So assume $\widetilde{k} \geq \ell$, in which case $\mathcal{S}$ is nonempty. Then

$$
\begin{aligned}
f(P) \;\geq\; & f(\widetilde{P}) \\
\geq\; & \max_{Q \in \mathcal{S}} f(Q) & (5.16) \\
\geq\; & \frac{1}{|\mathcal{S}|} \sum_{Q \in \mathcal{S}} f(Q) \\
=\; & \frac{1}{|\mathcal{S}|} \sum_{Q \in \mathcal{S}} \sum_{i} w_i \, \mathcal{A}_i(Q) \\
=\; & \frac{1}{|\mathcal{S}|} \sum_{Q \in \mathcal{S}} \sum_{i} \sum_{j \in \mathcal{O}_i(Q)} \frac{w_i \, a_{ij}}{|\mathcal{O}_i(Q)|} \\
=\; & \frac{1}{|\mathcal{S}|} \sum_{Q \in \mathcal{S}} \sum_{j \in Q} \sum_{i \,:\, j \in \mathcal{O}_i(Q)} \frac{w_i \, a_{ij}}{|\mathcal{O}_i(Q)|}\,, & (5.17)
\end{aligned}
$$

where inequality (5.16) holds because $\widetilde{P}$ is an optimal set of size at most ℓ and each Q is a set of size ℓ, while Eq. (5.17) just changes the order of summation on i and j.

Note that for any subset $Q \subseteq P^*$ and any fixed parameter choice $j \in Q$, the following relationship on sets of benchmarks holds:

$$
\bigl\{ i \,:\, j \in \mathcal{O}_i(P^*) \bigr\} \;\subseteq\; \bigl\{ i \,:\, j \in \mathcal{O}_i(Q) \bigr\}\,, \qquad (5.18)
$$

since if choice j is within tolerance ϵ of the highest estimator value for P^*, then j is within ϵ of the highest value for Q.

Continuing from Eq. (5.17), applying relationship (5.18) to index i of the innermost sum and observing that the terms lost are nonnegative, yields the following inequality (5.19):

$$f(P) \geq \frac{1}{|\mathcal{S}|} \sum_{Q \in \mathcal{S}} \sum_{j \in Q} \sum_{i : j \in \mathcal{O}_i(Q)} \frac{w_i\, a_{ij}}{|\mathcal{O}_i(Q)|}$$

$$\geq \frac{1}{|\mathcal{S}|} \sum_{Q \in \mathcal{S}} \sum_{j \in Q} \sum_{i : j \in \mathcal{O}_i(P^*)} \frac{w_i\, a_{ij}}{|\mathcal{O}_i(Q)|} . \tag{5.19}$$

Now define, for each benchmark i, a parameter choice $J(i)$ from P^* of highest estimator value,

$$J(i) \in \underset{j \in P^*}{\operatorname{argmax}} \left\{ E(A_{ij}) \right\},$$

where ties in the maximum estimator value are broken arbitrarily. Observe that when $J(i) \in Q$, the relationship $\mathcal{O}_i(Q) \subseteq \mathcal{O}_i(P^*)$ holds, since then both Q and P^* have the same highest estimator value (and $Q \subseteq P^*$). Thus when $J(i) \in Q$,

$$\left| \mathcal{O}_i(Q) \right| \leq \left| \mathcal{O}_i(P^*) \right| . \tag{5.20}$$

Returning to inequality (5.19), and applying relationship (5.20) in inequality (5.21) below,

$$f(P) \geq \frac{1}{|\mathcal{S}|} \sum_{Q \in \mathcal{S}} \sum_{j \in Q} \sum_{i : j \in \mathcal{O}_i(P^*)} \frac{w_i\, a_{ij}}{|\mathcal{O}_i(Q)|}$$

$$= \frac{1}{|\mathcal{S}|} \sum_{i} \sum_{Q \in \mathcal{S}} \sum_{j \in \mathcal{O}_i(P^*)} \frac{w_i\, a_{ij}}{|\mathcal{O}_i(Q)|}$$

$$\geq \frac{1}{|\mathcal{S}|} \sum_{i} \sum_{Q \in \mathcal{S} : J(i) \in Q} \sum_{j \in \mathcal{O}_i(P^*)} \frac{w_i\, a_{ij}}{|\mathcal{O}_i(Q)|}$$

$$\geq \frac{1}{|\mathcal{S}|} \sum_{i} \sum_{Q \in \mathcal{S} : J(i) \in Q} \sum_{j \in \mathcal{O}_i(P^*)} \frac{w_i\, a_{ij}}{|\mathcal{O}_i(P^*)|} \tag{5.21}$$

$$= \frac{1}{|\mathcal{S}|} \sum_{i} \left| \{ Q \in \mathcal{S} : J(i) \in Q \} \right| \sum_{j \in \mathcal{O}_i(P^*)} \frac{w_i\, a_{ij}}{|\mathcal{O}_i(P^*)|}$$

$$= \frac{\binom{\widetilde{k}-1}{\ell-1}}{\binom{\widetilde{k}}{\ell}} \sum_i \sum_{j \in \mathcal{O}_i(P^*)} \frac{w_i \, a_{ij}}{|\mathcal{O}_i(P^*)|}$$

$$= \left(\ell \big/ \widetilde{k}\right) f(P^*)$$

$$\geq \left(\ell / k\right) f(P^*).$$

Thus Greedy achieves approximation ratio at least ℓ/k.

Finally, to bound the running time of Greedy, consider an input instance with d features, n benchmarks, and m parameter choices in universe U. There are at most m^ℓ subsets of U of size at most ℓ, and evaluating objective function f on such a subset takes $O(d\ell n)$ time, so finding the optimal subset of size at most ℓ in the first step of Greedy takes $O(d\ell n m^\ell)$ time. The remaining for-loop considers at most k cardinalities, at most m parameter choices for each cardinality, and evaluates the objective function for each parameter choice on a subset of size at most k, which takes $O(dk^2mn)$ time. Thus the total time for Greedy is $O(d\ell n m^\ell + dk^2mn)$. For constant ℓ, this is polynomial time. □

In practice, optimal advisor sets of size up to $\ell = 5$ can be computed by exhaustive enumeration, as shown in Sect. 6.4.1. Finding an optimal advisor set of size $k = 10$, however, is currently far out of reach. Nevertheless, Theorem 5.1 shows we can still find reasonable approximations even for such large advisor sets, since for $\ell = 5$ and $k = 10$, Greedy is a $(1/2)$-approximation algorithm.

We next show it is not possible to prove a greater approximation ratio than in Theorem 5.1, as that ratio is in fact tight.

Theorem 5.2 (Tightness of Approximation Ratio) *The approximation ratio ℓ/k for algorithm* Greedy *is tight.*

Proof. Since the ratio is obviously tight for $\ell = k$, assume $\ell < k$. For any arbitrary constant $0 < \delta < 1-(\ell/k)$, and for any error tolerance $0 \leq \epsilon < 1$, consider the following infinite class of instances of Advisor Set with:

- benchmarks $1, 2, \ldots, n$,
- benchmark weights $w_i = 1/n$,
- cardinality bound $k = n$, and
- universe $U = \{0, 1, \ldots, n\}$ of $n+1$ parameter choices.

The estimator values for all benchmarks i are,

$$E(A_{ij}) = \begin{cases} 1, & j = 0; \\ (1-\epsilon)/2, & i = j > 0; \\ 0, & \text{otherwise}; \end{cases}$$

which can be achieved by appropriate feature vectors F_{ij}. The alternate alignment accuracies for all benchmarks i are,

$$a_{ij} = \begin{cases} (\ell/k) + \delta, & j = 0; \\ 1, & i = j > 0; \\ 0, & \text{otherwise.} \end{cases}$$

For such an instance of Advisor Set, an optimal set of size at most k is $P^* = \{1, \ldots, n\}$, which achieves $f(P^*) = 1$. Every optimal set $\widetilde{P}$ of size at most $\ell < k$ satisfies $\widetilde{P} \supseteq \{0\}$: it cannot include all of parameter choices $1, 2, \ldots, n$, so to avoid getting accuracy 0 on a benchmark it must contain parameter choice $j = 0$. Moreover, every such set $\widetilde{P} \supseteq \{0\}$ has average accuracy $f(\widetilde{P}) = (\ell/k) + \delta$: parameter choice $j = 0$ has the maximum estimator value 1 on every benchmark, and no other parameter choice $j \neq 0$ has estimator value within ϵ of the maximum, so on every benchmark $\mathcal{A}_i(\widetilde{P}) = (\ell/k) + \delta$. Furthermore, every greedy augmentation $P \supseteq \widetilde{P}$ also has this same average accuracy $f(P) = f(\widetilde{P})$. Thus on this instance the advisor set P output by Greedy has approximation ratio exactly

$$\frac{f(P)}{f(P^*)} = \frac{\ell}{k} + \delta.$$

Now suppose the approximation ratio from Theorem 5.1 is not tight, in other words, that an even better approximation ratio $\alpha > \ell/k$ holds. Then take $\delta = (\alpha - (\ell/k))/2$, and run Greedy on the above input instance. On this instance, Greedy only achieves ratio

$$\frac{\ell}{k} + \delta = \frac{1}{2}\left(\frac{\ell}{k} + \alpha\right) < \alpha,$$

a contradiction. So the approximation ratio is tight.   □

Summary

This chapter presented integer linear programs for finding optimal advisor sets, and more generally, for finding an optimal parameter advisor. As these integer programs are not solvable in practice, we also developed an efficient approximation algorithm for finding estimator-aware advisor sets. As shown in Chap. 6, experiments with an implementation of the approximation algorithm on biological benchmarks, using various accuracy estimators from the literature, demonstrate it finds advisor sets that are surprisingly close to optimal. Furthermore, the resulting parameter advisors are significantly more accurate than simply aligning with a single default parameter choice.

Part II
Applications of Parameter Advising

Chapter 6
Parameter Advising for the `Opal` Aligner

Chapters 1 through 5 described several approaches to constructing a parameter advisor. This chapter demonstrates the performance of the resulting advisors, learned for the `Opal` aligner, trained on a suite of benchmark reference alignments. Advising performance is compared against the optimal default parameter choice, as well as advisors learned using various accuracy estimators. The results show that `Facet` yields the best advising accuracy of any estimator currently available, and that by using estimator-aware advisor sets we can significantly increase advising accuracy over using estimator-oblivious oracle sets.

Looking Back

Recall from earlier chapters that the task of *parameter advising* is as follows: given particular sequences to align, and a set of possible parameter choices, recommend a parameter choice to the aligner that yields the most accurate alignment of those sequences. Parameter advising has three components: the set S of input sequences, the set P of parameter choices, and the aligner $\mathcal{A}$. (Here a *parameter choice* $p \in P$ is a vector $p = (p_1, \ldots, p_k)$ that specifies values for *all* free parameters in the alignment scoring function.) Given sequences S and parameter choice $p \in P$, we denote the alignment output by the aligner as $\mathcal{A}_p(S)$. A procedure that takes the set of input sequences S and the set of parameter choices P, and outputs a parameter recommendation $p \in P$, an *advisor*. A perfect advisor, that always recommends the choice $p^* \in P$ that yields the highest accuracy alignment $\mathcal{A}_{p^*}(S)$, is called an *oracle*. In practice, constructing an oracle is impossible, since for any real set S of sequences that we want to align, a reference alignment for S is unknown (as otherwise we

---

Adapted from publications [24, 26, 33, 56].

© Springer International Publishing AG 2017

D. DeBlasio, J. Kececioglu, *Parameter Advising for Multiple Sequence Alignment*,
Computational Biology 26, https://doi.org/10.1007/978-3-319-64918-4_6

would not need to align them), so the true accuracy of any alignment of S cannot be determined. The concept of an oracle is useful, however, for measuring how well an actual advisor performs.

A natural approach for constructing a parameter advisor is to use an accuracy estimator E as a proxy for true accuracy, and recommend the parameter choice

$$\widetilde{p} := \underset{p \in P}{\mathrm{argmax}}\ E\big(\mathcal{A}_p(S)\big).$$

In its simplest realization, such an advisor will run the aligner $\mathcal{A}$ repeatedly on input S, once for each possible parameter choice $p \in P$, to select the output that has best estimated accuracy. Of course, to yield a quality advisor, this requires two ingredients: a good estimator E, and a good set P of parameter choices.

In Chaps. 2 and 3 we presented a framework for accuracy estimation that lead to the new accuracy estimator Facet (short for "feature-based accuracy estimator") which is a linear combination of easy-to-compute feature functions of an alignment. Chapter 5 went on to present a greedy approximation algorithm for finding advisor sets. Note that as discussed in Chap. 4, finding optimal advisor sets is NP-complete.

Combining the results of the previous chapters provides the means to compute both accuracy estimators and advisor sets. The rest of this chapter applies all of this methodology to the task of parameter advising.

6.1   Experimental Methods

The approach for deriving an accuracy estimator, and the quality of the resulting parameter advisor are evaluated through experiments on a collection of benchmark protein multiple sequence alignments. These experiments compare parameter advisors that use the Facet estimator and five other estimators from the literature: COFFEE [77], NorMD [99], MOS [67], HoT [64], and PredSP [2]. (In terms of our earlier categorization of estimators, COFFEE, NorMD and PredSP are scoring-function-based, while MOS and HoT are support-based.) Other estimators from the literature that are not in this comparison group are: AL2CO [81], which is known to be dominated by NorMD [67] and PSAR [61], which at present is only implemented for DNA sequence alignments.

The collection of alignment benchmarks used in the following experiments include the BENCH suite of [39], which consists of 759 benchmarks, supplemented by a selection of 102 benchmarks from the PALI suite of [9]. (BENCH itself is a selection of 759 benchmarks from [8], OxBench [87], and SABRE [102].) Both BENCH and PALI consist of protein multiple sequence alignments mainly induced by structural alignment of the known three-dimensional structures of the proteins. The entire benchmark collection consists of 861 reference alignments.

For the experiments the *difficulty* of a benchmark S is defined by the true accuracy of the alignment computed by the multiple alignment tool Opal [106, 107] on sequences S using its default parameter choice, where the computed alignment is compared to the benchmark's reference alignment on its core columns. Using this measure, the 861 benchmarks were binned by difficulty, where the full range $[0, 1]$ of accuracies were divided into 10 bins with difficulties $[(i - 1)/10, i/10]$ for $i = 1, \ldots, 10$. As is common in benchmark suites, easy benchmarks are highly over-represented compared to hard benchmarks. The number of benchmarks falling in bins $[0.0, 0.1]$ through $[0.9, 1.0]$ are listed below.

Bin	0.1	0.2	0.3	0.4	0.5	0.6	0.7	0.8	0.9	1.0
Benchmarks	12	12	20	34	26	50	61	74	137	434

As was mentioned in previous chapters, to correct for this bias in oversampling of easy benchmarks, the approaches described for learning an estimator nonuniformly weight the training examples.

Notice that with this uniform weighting of bins, the singleton advising set P containing *only* the optimal default parameter choice will tend to an average advising accuracy $f(P)$ of 50% (illustrated later in Figs. 6.2 and 6.3). This establishes, as a point of reference, average accuracy 50% as the *baseline* against which to compare advising performance.

Note that if advising accuracy was instead measured by uniformly averaging over *benchmarks*, then the predominance of easy benchmarks (for which little improvement is possible over the default parameter choice) makes both good and bad advisors tend to an average accuracy of nearly 100%. Uniformly averaging over *bins* allows for discrimination among advisors, though the average advising accuracies are now pulled down from 100% toward 50%.

For each reference alignment in the benchmark collection, alternate multiple alignments of the sequences in the reference were generated using Opal with varying parameter choices. Opal constructs multiple sequence alignments using as a building block the exact algorithm of [58] for optimally aligning two multiple alignments under the sum-of-pairs scoring function [16] with affine gap penalties [45]. Since Opal computes subalignments that are optimal with respect to a well-defined scoring function, it is an ideal testbed for evaluating parameter choices, and in particular parameter advising. Each *parameter choice* for Opal is a five-tuple $(\sigma, \gamma_I, \gamma_E, \lambda_I, \lambda_E)$ of parameter values, where σ specifies the amino acid substitution scoring matrix, pair γ_E, λ_E specifies the gap-open and gap-extension penalties for *external* gaps in the alignment (also called terminal gaps), and γ_I, λ_I specifies the gap penalties for *internal* gaps (or non-terminal gaps).

The universe U of parameter choices considered in the experiments consists of over 2000 such tuples $(\sigma, \gamma_I, \gamma_E, \lambda_I, \lambda_E)$. Universe U was generated as follows. For the substitution matrix σ, several matrices from the BLOSUM [48] and VTML [74] families were considered. To accommodate a range of protein sequence divergences, the following matrices from these families are included: {BLSM45, BLSM62, BLSM80} and {VTML20, VTML40, VTML80, VTML120, VTML200}. For each of these eight matrices, the real-valued version of the similarity matrix was transformed it into a substitution cost matrix for Opal by negating, shifting, and scaling it to the range [0, 100], and then rounding its entries to the nearest integer. For the gap penalties, the default parameter setting for Opal (see [106]) was used as a starting point, this default is an optimal choice of gap penalties for the BLSM62 matrix found by inverse parametric alignment (see [57, 60].) Around these default values a Cartesian product of integer choices in the neighborhood of this central choice was enumerated, generating over 2100 four-tuples of gap penalties. The resulting set of roughly 16,900 parameter choices (each substitution matrix combined with each gap penalty assignment) was then reduced by examining the benchmarks in our collection as follows. In each hardness bin of benchmarks: (1) run Opal with all of these parameter choices on the benchmarks in the bin, (2) for a given parameter choice measure the average accuracy of the alignments computed by Opal using that parameter choice on the bin, (3) sort the parameter choices for a bin by their average accuracy, and (4) in each bin kept the top 25 choices with highest average accuracy. Unioning these top choices from all 10 hardness bins, and removing duplicates, gave the final set U. This universe U has 243 parameter choices.

To generate training and testing sets for the experiments on learning advisor sets, 12-*fold cross validation* was used. For each hardness bin, the benchmarks in the bin were evenly and randomly partitioned into twelve groups; twelve splits of the entire collection of benchmarks were formed into a training class and a testing class, where each split placed one group in a bin into the testing class and the other eleven groups in the bin into the training class; finally, a *training set* and a *testing set* of example alignments is generated as follows: for each benchmark B in a training or testing class, generate $|U|$ example alignments in the respective training or testing set by running Opal on B with each parameter choice from U. An estimator learned on the examples from a training set was evaluated on examples from the corresponding testing set. The results reported later are averages over twelve folds, where each *fold* is one of these pairs of associated training and testing sets. (Note that across the twelve folds, every example is tested on exactly once.) Each fold contains over 190,000 training examples.

To compare with the GUIDANCE accuracy estimator a smaller set of benchmarks was used. GUIDANCE requires each benchmark to contain at least four sequences, a requirement that is not met by all of the benchmarks in the collection described above. The subset of benchmarks used later only for selected results includes the subset of the universe which contains four sequences, more precisely a set of 605 benchmarks. These benchmarks are split into the same bins as defined earlier with a similar frequency of bin sizes.

Bin	0.1	0.2	0.3	0.4	0.5	0.6	0.7	0.8	0.9	1.0
Benchmarks	4	10	9	16	21	30	40	56	96	323

Due to the reduced number of benchmarks in the smaller bins, when evaluating the GUIDANCE estimator, 4-fold cross validation was used on the reduced benchmark collection described earlier, with folds generated by the above procedure. Each of these folds still contains over 109,000 training examples.

6.2   Comparing Accuracy Estimators

The previous section described the full experimental setup for the results of parameter advising described below. The next sections show the trained Facet estimator, how it compares with other accuracy estimators, and use the Facet estimator to find greedy advisor sets. The learned estimator and advisor sets are then applied to parameter advising with the Opal aligner. The results show that Facet is the most accurate estimator in terms of parameter advising accuracy, and that the accuracy of advisors using greedy advisor sets are higher than those using oracle sets.

6.2.1   Learning the Facet Estimator

Coefficients for the Facet estimator were found using the difference-fitting method described in Sect. 2.2.2. The threshold-difference pairs method was used with $\epsilon = 5\%$, for all 16,896 realignments of each benchmark ($\ell = |U| = 243$ and $k = 16,896$). Note that here an estimator is learned for pairs from all 861 benchmarks, as opposed to just one of the training folds. The estimators evaluated in later experiments involving parameter advising use cross-validation training sets to learn new estimator coefficients for each fold, which avoids testing on benchmarks that were used for training the estimator or advisor sets.

Of the features listed in Sect. 3.1, not all are equally informative, and some can weaken an estimator. When coefficients are found by solving the linear programs described in Chap. 2 on a set of example alignments, some of the coefficients of the estimator will be zero. The best overall feature set found by this process is a six-feature subset consisting of the following feature functions:

- Secondary Structure Agreement, f_{SA},
- Secondary Structure Blockiness, f_{BL},
- Secondary Structure Identity, f_{SI},
- Gap Open Density, f_{GO},
- Amino Acid Identity, f_{AI}, and
- Core Column Percentage, f_{CC}.

The corresponding fitted estimator is

$$E(A) = 0.239 f_{SA}(A) + 0.141 f_{BL}(A) + 0.040 f_{SI}(A) +$$
$$0.465 f_{GO}(A) + 0.204 f_{AI}(A) + 0.003 f_{CC}(A).$$

Figure 3.1 shows a scatter plot of the five strongest features from the estimator. Notice that the feature with the highest coefficient value also has the smallest *range*.

If instead the distributed-example pairs method is used, a much different estimator is found. The features used are

- Secondary Structure Agreement, f_{SA},
- Secondary Structure Blockiness, f_{BL},
- Secondary Structure Identity, f_{SI},
- Substitution Compatibility, f_{SP}, and
- Average Substitution Score, f_{AS}.

and the learned estimator is

$$E(A) \;=\; 0.174\, f_{SA}(A) \;+\; 0.172\, f_{BL}(A) \;+\; 0.168\, f_{SI}(A) \;+$$
$$0.167\, f_{SP}(A) \;+\; 0.152\, f_{AS}(A) \;+\; 0.167.$$

When using distributed-example pairs, the LP is unable to perform parameter selection, so this must be done manually. To determine the estimator, all 4095 subsets of the 12 features were enumerated. To fit an estimator on a feature set, both value fitting and difference fitting were used, for both linear and quadratic estimators (i.e. degree 1 and 2 polynomials), under both the L_1 norm and L_2 norm error measures. Each of the resulting estimators was evaluated for its performance on parameter advising, in terms of the true accuracy of its resulting advisor averaged across the ten difficulty bins. This evaluation process was used to find good feature subsets for the optimal parameter sets $P \subseteq U$ with $|P|$ equal to 5, 10, and 15 parameter choices. To find a good *overall* feature set that works well for differing numbers of parameters, we examined all subsets of features considered by the above process, and chose the feature set that had the highest accuracy averaged across the 5-, 10-, and 15-parameter sets. The coefficients shown above are an average of the corresponding coefficients from the estimators resulting from fitting on training sets for the cross-validation folds for each of the above three parameter sets. This linear estimator was obtained using difference fitting under the L_1 norm.

In the experiments that follow in this chapter, we use distributed-example pairs, and the feature subset listed above, to learn the estimator by difference fitting.

6.2.2   Comparing Estimators to True Accuracy

To examine the fit of an estimator to true accuracy, the scatter plots in Fig. 6.1 show the value of an estimator versus true accuracy on all example alignments in the 15-parameter test set. (This set has over 12,900 test examples. Note that these test examples are disjoint from the training examples used to fit our estimator.) The scatter plots show our Facet estimator, as well as the PredSP, MOS, COFFEE, HoT, and NorMD estimators. Note that the MOS estimator, in distinction to the other

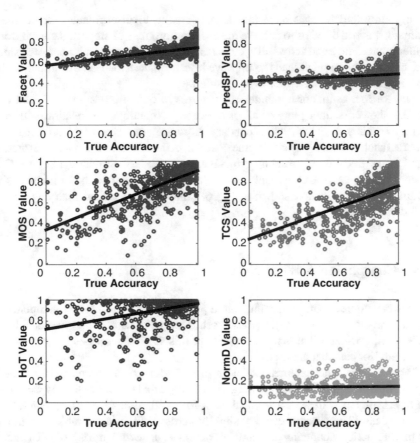

Fig. 6.1 Correlation of estimators with accuracy. Each scatter plot shows the value of an estimator versus true accuracy for alignments of the 861 benchmarks used for testing aligned with the default parameter settings for the Opal aligner.

estimators, receives as input *all* the alternate alignments of an example's sequences generated by the 15 parameter choices, which is much more information than is provided to the other estimators, which are only given the *one* example alignment.

An ideal estimator would be monotonic increasing in true accuracy. A real estimator approaches this ideal according to its *slope* and *spread*. To discriminate between low and high accuracy alignments for parameter advising, an estimator needs large slope with small spread. Comparing the scatter plots by spread, Facet and PredSP have the smallest spread; MOS and COFFEE have intermediate spread; and HoT and NorMD have the largest spread. Comparing by slope, PredSP and NorMD have the smallest slope; Facet and HoT have intermediate slope; and MOS and COFFEE have the largest slope. While PredSP has small spread, it also has small slope, which weakens its discriminative power. While MOS and COFFEE have large slope, they also have significant spread, weakening their discrimination.

Finally HoT and NorMD have too large a spread to discriminate. Of all these estimators, Facet seems to achieve the best compromise of slope and spread, for a tighter monotonic trend across all accuracies. This better compromise between slope and spread may be what leads to improved performance for Facet on parameter advising, as demonstrated later in this section.

The Facet estimator combines six features to obtain its estimate. Recall that Fig. 3.1 showed scatter plots of all of the feature functions' correlation with true accuracy (many of which use secondary structure). As noted in Sect. 6.2.1, the feature functions that we use for the Facet estimator are: Secondary Structure Agreement, Secondary Structure Blockiness, Secondary Structure Identity, Gap Open Density, and Average Substitution Score. Notice that the combined six-feature Facet estimator, shown in Fig. 6.1, has smaller spread than any one of its individual features.

6.3   Comparing Advisor Sets

As has been noted many times thus far, a parameter advisor has two components: the accuracy estimator and the advisor set. The previous section gave details on the Facet estimator that is used in practice; this section describes the greedy and oracle advisor sets that are used.

Table 6.1 lists the parameter choices in the advisor sets found by the *greedy* approximation algorithm (augmenting from the optimal set of cardinality $\ell = 1$) for the Opal aligner with the Facet estimator for cardinalities $k \leq 20$, on one fold of training data. (The greedy sets vary slightly across folds.) In the table, the greedy set of cardinality k contains the parameter choices at rows 1 through k. (The entry at row 1 is the optimal *default parameter choice*.) As we noted earlier a parameter choice for the Opal aligner is five-tuple $(\sigma, \gamma_I, \gamma_E, \lambda_I, \lambda_E)$, where γ_I and γ_E are gap-open penalties for non-terminal and terminal gaps respectively, and λ_I and λ_E are corresponding gap-extension penalties. The scores in the substitution matrix σ are dissimilarity values scaled to integers in the range [0, 100]. (The associated gap penalty values in a parameter choice relate to this range.) The accuracy column gives the average advising accuracy (in Opal using Facet) of the greedy set of cardinality k on *training* data, uniformly averaged over benchmark bins. Recall this averaging will tend to yield accuracies close to 50%.

Interestingly, while BLOSUM62 [48] is the substitution scoring matrix most commonly used by standard aligners, it does not appear in a greedy set until cardinality $k = 11$. The VTML family [74] appears more often than BLOSUM. A plateau in advising accuracy, which will be described in detail in later sections, is evident in this single training instance, though ever more gradual improvement remains as cardinality k increases.

Table 6.1 Greedy advisor sets for Opal using Facet

Cardinality k	Parameter choice $(\sigma, \gamma_I, \gamma_E, \lambda_I, \lambda_E)$	Average advising accuracy (%)
1	(VTML200, 50, 17, 41, 40)	51.2
2	(VTML200, 55, 30, 45, 42)	53.4
3	(BLSUM80, 60, 26, 43, 43)	54.5
4	(VTML200, 60, 15, 41, 40)	55.2
5	(VTML200, 55, 30, 41, 40)	55.6
6	(BLSUM45, 65, 3, 44, 43)	56.1
7	(VTML120, 50, 12, 42, 39)	56.3
8	(BLSUM45, 65, 35, 44, 44)	56.5
9	(VTML200, 45, 6, 41, 40)	56.6
10	(VTML120, 55, 8, 40, 37)	56.7
11	(BLSUM62, 80, 51, 43, 43)	56.8
12	(VTML120, 50, 2, 45, 44)	56.9
13	(VTML200, 45, 6, 40, 40)	57.0
14	(VTML40, 50, 2, 40, 40)	57.1
15	(VTML200, 50, 12, 43, 40)	57.2
16	(VTML200, 45, 11, 42, 40)	57.3
17	(VTML120, 60, 9, 40, 39)	57.3
18	(VTML40, 50, 17, 40, 38)	57.4
19	(BLSUM80, 70, 17, 42, 41)	57.4
20	(BLSUM80, 60, 3, 42, 42)	57.6

6.3.1   Shared Structure Across Advisor Sets

To assess the similarity of advisor sets found by the three approaches considered
in our experiments—*greedy sets* via the approximation algorithm, *exact sets* via
exhaustive search, and *oracle sets* via integer linear programming—the overlap both
within and between folds should be examined.

Table 6.2 shows the composition of the greedy, exact, and oracle sets for the
training instance in one fold, at cardinality $k = 2, 3, 4$ and greedy set finding
tolerance $\epsilon = 0$. A non-blank entry in the table indicates that the parameter choice at
its row is contained in the advisor set at its column. (The column labeled "default"
indicates the optimal *default parameter choice* for the fold, or equivalently, the
exact set of cardinality $k = 1$.) The value in parentheses at an entry is the number
of folds (for twelve-fold cross-validation) where that parameter choice appears in
that advisor set. (For example, at cardinality $k = 4$, the second parameter choice
(VTML200, 55, 30, 45, 42) is in the greedy, exact, and oracle sets for this particular
fold, and overall is in exact sets for 9 of 12 folds, including this fold.) Surprisingly,
the default parameter choice (the best single choice) never appears in the exact or
oracle sets for this fold at any of the cardinalities beyond $k = 1$, and also is reused as

Table 6.2 Composition of advisor sets at different cardinalities k

Parameter choice $(\sigma, \gamma_I, \gamma_E, \lambda_I, \lambda_E)$	Default	Advisor set Greedy	Exact	Oracle
$k = 2$				
(VTML200, 50, 17, 41, 40)	(2)	(2)		
(VTML200, 55, 30, 45, 42)		(2)	(3)	(1)
(BLSUM80, 60, 9, 43, 42)			(2)	
(BLSUM45, 65, 35, 44, 44)				(3)
$k = 3$				
(VTML200, 50, 17, 41, 40)	(2)	(2)		
(VTML200, 55, 30, 45, 42)		(3)	(5)	(1)
(BLSUM80, 60, 26, 43, 43)		(2)	(2)	
(VTML200, 55, 30, 41, 40)			(6)	
(VTML40, 45, 29, 40, 39)				(7)
(BLSUM62, 65, 16, 44, 42)				(8)
$k = 4$				
(VTML200, 50, 17, 41, 40)	(2)	(2)		
(VTML200, 55, 30, 45, 42)		(3)	(9)	(6)
(BLSUM80, 60, 26, 43, 43)		(2)		
(VTML200, 60, 15, 41, 40)		(1)		
(VTML200, 45, 6, 40, 40)			(8)	(1)
(VTML200, 55, 30, 41, 40)			(8)	
(BLSUM80, 55, 19, 43, 42)			(1)	
(VTML40, 45, 29, 40, 39)				(4)
(BLSUM62, 65, 35, 44, 42)				(3)

Table 6.3 Number of folds where greedy and exact sets share parameters

Intersection cardinality	Advisor set cardinality $k = 2$	$k = 3$	$k = 4$	$k = 5$
0	9	4	3	2
1	3	5	6	5
2	0	3	3	4
3		0	0	1
4			0	0
5				0

the default in only one other fold. In general there is relatively little overlap between these advisor sets: often just one and at most two parameter choices are shared.

Table 6.3 examines whether this trend continues at other folds, by counting how many training instances (out of the twelve folds) share a specified number of parameter choices between their *greedy* and *exact* sets, for a given advisor set cardinality k. (For example, at cardinality $k = 4$, six training instances share exactly one parameter choice between their greedy and exact sets; in fact, the fold shown in Table 6.2 is one such instance.) On the whole, the two "estimator-aware" advisor sets—the greedy and exact sets—are relatively dissimilar, and never share more than $\lceil k/2 \rceil$ parameter choices.

6.4 Application to Parameter Advising

Given the accuracy estimator learned using difference fitting described in earlier sections, and the advisor sets described in the previous section, we now evaluate the advising accuracy of the resulting parameter advisor.

6.4.1 Learning Advisor Sets by Different Approaches

The advising accuracy of parameter sets learned for the Facet estimator by different approaches is examined first. The protocol used first constructs an optimal *oracle* set for cardinalities $1 \leq k \leq 20$ for each training instance. A coefficient vector for the advisor's estimator was then found for each of these oracle sets by the difference-fitting method. Using this estimator learned for the training data, exhaustive search was done to find optimal *exact* advisor sets for cardinalities $k \leq 5$. The optimal exact set of size $\ell = 1$ (the best default parameter choice) was then used as the starting point to find near-optimal *greedy* advisor sets by our approximation algorithm for $k \leq 20$. Each of these advisors (an advising set combined with the estimator) was then used for parameter advising in Opal, returning the computed alignment with highest estimator value. These set-finding approaches are compared based on the accuracy of the alignment chosen by the advisor, averaged across bins (Figs. 6.2 and 6.3).

Figure 6.4 shows the performance of these advisor sets under twelve-fold cross validation. The left plot shows advising accuracy on the testing data averaged over the folds, while the right plot shows this on the training data.

Notice that while there is a drop in accuracy when an advising set learned using the greedy and exact methods is applied to the testing data, the drop in accuracy is greatest for the exact sets. The value of ϵ shown in the plot maximizes the accuracy of the resulting advisor on the testing data. Notice also that for cardinality $k \leq 5$ (for which exact sets could be computed), on the testing data the greedy sets are often performing as well as the optimal exact sets.

Figures 6.2 and 6.3 show the performance within each benchmark bin when advising with Facet using greedy sets of cardinality $k = 5, 10, 15$ ($k = 5$ and 10 in Fig. 6.2 top and bottom respectively, $k = 15$ in Fig. 6.3) Notice that for many bins, the performance is close to the best-possible accuracy attainable by any advisor, shown by the dashed line for a perfect oracle advisor. The greatest boost over the default parameter choice is achieved on the bottom bins that contain the hardest benchmarks.

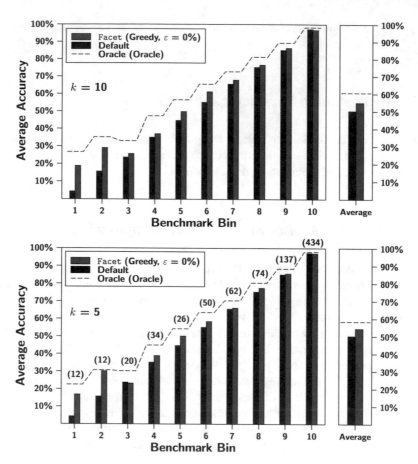

Fig. 6.2 Advising accuracy of Facet within benchmark bins. These bar charts show the advising accuracy of various approaches to finding advisor sets, for cardinality $k = 5, 10$. For each cardinality, the horizontal axis of the chart on the left corresponds to benchmark bins, and the vertical bars show advising accuracy averaged over the benchmarks in each bin. Black bars give the accuracy of the optimal *default* parameter choice, and red bars give the accuracy of advising with Facet using the *greedy* set. The dashed line shows the limiting performance of a perfect advisor: an oracle with true accuracy as its estimator using an optimal *oracle* set. In the top chart, the numbers in parentheses above the bars are the number of benchmarks in each bin. The narrow bar charts on the right show advising accuracy uniformly averaged over the bins.

6.4.2   Varying the Exact Set for the Greedy Algorithm

The advising accuracy of the greedy sets learned when using cardinalities $1 \le \ell \le 5$ were examined to find the appropriate cardinality ℓ of the initial exact solution that is augmented within approximation algorithm Greedy. Figure 6.5 shows the accuracy of the resulting advisor using greedy sets of cardinality $1 \le k \le 20$,

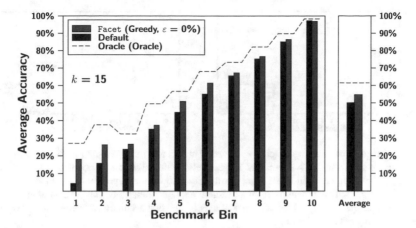

Fig. 6.3 Advising accuracy of Facet within benchmark bins. These bar charts show the advising accuracy of various approaches to finding advisor sets, for cardinality $k = 15$. For each cardinality, the horizontal axis of the chart on the left corresponds to benchmark bins, and the vertical bars show advising accuracy averaged over the benchmarks in each bin. Black bars give the accuracy of the optimal *default* parameter choice, and red bars give the accuracy of advising with Facet using the *greedy* set. The dashed line shows the limiting performance of a perfect advisor: an oracle with true accuracy as its estimator using an optimal *oracle* set. In the top chart, the numbers in parentheses above the bars are the number of benchmarks in each bin. The narrow bar charts on the right show advising accuracy uniformly averaged over the bins.

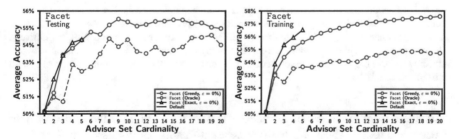

Fig. 6.4 Advising using exact, greedy, and oracle sets with Facet. The plots show advising accuracy using the Facet estimator with parameter sets learned by the optimal *exact* algorithm and the *greedy* approximation algorithm for Advisor Set, and with *oracle* sets. The horizontal axis is the cardinality of the advisor set, while the vertical axis is the advising accuracy averaged over the benchmarks. Exact sets are known only for cardinalities $k \leq 5$; greedy sets are augmented from the exact set of cardinality $\ell = 1$. The left and right plots show accuracy on the testing and training data, respectively, where accuracies are averaged over all testing or training folds.

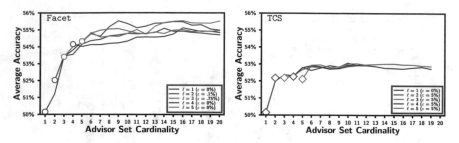

Fig. 6.5 Greedily augmenting exact advisor sets. The left and right plots show advising accuracy using the Facet and TCS estimators respectively, with advisor sets learned by procedure Greedy, which augments an exact set of cardinality ℓ to form a larger set of cardinality $k > \ell$. Each curve is greedily augmenting from a different exact cardinality ℓ. The horizontal axis is the cardinality k of the augmented set; the vertical axis is advising accuracy on testing data, averaged over all benchmarks and all folds.

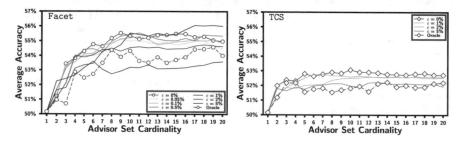

Fig. 6.6 Effect of error tolerance on advising accuracy using greedy sets. The plots show advising accuracy on testing data using greedy sets learned for the two best estimators, Facet and TCS, at various error tolerances $\epsilon \geq 0$. The plots on the left and right are for Facet and TCS, respectively. For comparison, both plots also include a curve showing performance using the estimator on oracle sets, drawn with a dashed line. The solid curves with circles and diamonds highlight the best overall error tolerance of $\epsilon = 0$.

augmented from exact sets of cardinality $1 \leq \ell \leq 5$, using for the estimator both Facet and TCS. (These are the two best estimators, as discussed in Sect. 6.4.3 below). The points plotted with circles show the accuracy of the optimal exact set that is used within procedure Greedy for augmentation (Fig. 6.6).

Notice that the initial exact set size ℓ has relatively little effect on the accuracy of the resulting advisor; at most cardinalities, starting from the single best parameter choice ($\ell = 1$) has highest advising accuracy. This is likely due to the behavior observed earlier in Fig. 6.4, namely that exact sets do not generalize as well as greedy sets.

6.4.3 Learning Advisor Sets for Different Estimators

In addition to learning advisor sets for Facet [56], advisor sets were also learned
for the best accuracy estimators from the literature: namely, TCS [18], MOS [67],
PredSP [2], and GUIDANCE [85]. The scoring-function-based accuracy estimators
TCS, PredSP, and GUIDANCE do not have any dependence on the advisor set
cardinality or the training benchmarks used. The support-based estimator MOS,
however, requires a set of alternate alignments in order to compute its estimator
value on an alignment. In each experiment, an alignment's MOS value was computed
using alternate alignments generated by aligning under the parameter choices in
the oracle set; if the parameter choice being tested on was in the oracle set, it was
removed from this collection of alternate alignments.

After computing the values of these estimators, exhaustive search was used to
find optimal exact sets of cardinality $\ell \leq 5$ for each estimator, as well as greedy sets
of cardinality $k \leq 20$ (augmenting from the exact set for $\ell = 1$).

The tendency of exact advisor sets to not generalize well is even more pronounced when accuracy estimators other than Facet are used. Figure 6.7 shows
the performance on testing and training data of greedy, exact, and oracle advisor sets
learned for the best three other estimators: TCS, MOS, and PredSP. The results for
greedy advisor sets for TCS at cardinalities larger than 5 have similar trend to those
seen for Facet (with now a roughly 1% accuracy improvement over the oracle
set), but surprisingly with TCS its exact set always has lower testing accuracy than
its greedy set. Interestingly, for MOS its exact set rarely has better advising accuracy
than the oracle set. For PredSP, at most cardinalities (with the exception of $k = 3$)
the exact set has higher accuracy than the greedy set on testing data, though this is
offset by the low accuracy of the estimator.

The reduced set of benchmarks described earlier that contain at least four
sequences was used to test the usefulness of generating advisor sets for the
GUIDANCE, Facet, and TCS estimators. Figure 6.8 shows the advising accuracy
of set-finding methods using these estimators on these benchmarks. Notice that on
this reduced suite the results generally stay the same, though for Facet there is
more of a drop in performance of the exact set from training to testing, and the set
found by Greedy generally has greater accuracy on the reduced suite than the full
suite.

Finally, a complete comparison of the advising performance of *all estimators*
using greedy sets is shown in Fig. 6.9. (The plot on the right shows advising
accuracy on testing data for GUIDANCE, Facet, and TCS on the reduced suite
of benchmarks with at least four sequences.) Advising with each of these estimators
tends to eventually reach an accuracy plateau, though their performance is always
boosted by using advisor sets larger than a singleton default choice. The plateau for
Facet (the top curve in the plots) generally occurs at the greatest cardinality and
accuracy.

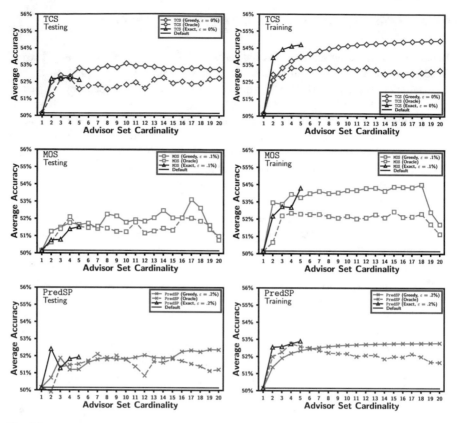

Fig. 6.7 Comparing testing and training accuracies of various estimators. The plots show the advising accuracies on testing and training data using TCS, MOS, and PredSP with parameter sets learned for these estimators by the *exact* and *greedy* algorithms for Advisor Set, and with *oracle* sets. From top to bottom, the estimators used are TCS, MOS, and PredSP, with *testing* data plotted on the left, and *training* data on the right.

6.4.4   Varying the Error Tolerance for the Greedy Algorithm

When showing experimental results, an error tolerance ϵ has always been used that yields the most accurate advisor on the testing data. Figure 6.6 shows the effect of different values of ϵ on the testing accuracy of an advisor using greedy sets learned for the Facet and TCS estimators. (While the same values of ϵ were tried for both estimators, raw TCS scores are integers in the range $[0, 100]$ which were scaled to real values in the range $[0, 1]$, so for TCS any $\epsilon < 0.1$ is equivalent to $\epsilon = 0$.) Conceptually, using a nonzero error tolerance $\epsilon > 0$ in the Greedy algorithm would be expected to boost the generalization of the advisor sets found. As this experiment shows, there seems to be no clear relationship between testing accuracy and error tolerance, though for Facet and TCS alike, setting $\epsilon = 0$ generally gives the best overall advising accuracy.

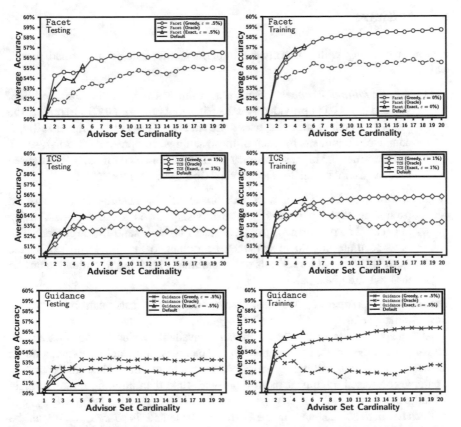

Fig. 6.8 Comparing testing and training accuracies of estimators on benchmarks with at least four sequences. The plots show advising accuracies for *testing* and *training* data on benchmarks with at least four sequences, using `Facet`, `TCS`, and `GUIDANCE` with *exact*, *greedy*, and *oracle* sets.

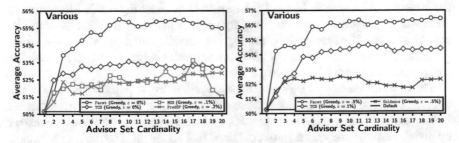

Fig. 6.9 Comparing all estimators on greedy advisor sets. The plots show advising accuracy on *greedy* sets learned for the estimators `Facet`, `TCS`, `MOS`, `PredSP`, and `GUIDANCE`. The vertical axis is advising accuracy on *testing* data, averaged over all benchmarks and all folds. The horizontal axis is the cardinality k of the greedy advisor set. Greedy sets are augmented from the exact set of cardinality $\ell = 1$. The plot on the left uses the *full* benchmark suite; the plot on the right, which includes `GUIDANCE`, uses a *reduced* suite of all benchmarks with at least four sequences.

6.5 Software

Parameter advising with our software implementation can be performed in two ways:

(1) ***Facet aligner wrapper***   Similar to using Facet on the command line, you can use a set of provided Perl scripts that runs PSIPRED to predict the protein secondary structure, uses a provided set of Opal parameter settings, computes alignments for each of these settings, computes the Facet score, and identifies the highest accuracy alignment. The script must be configured for each user's installation location of Opal and PSIPRED.

(2) ***Within Opal***   The newest version of the Opal aligner can perform parameter advising internally. The advising set is given to Opal using the -advising_configuration_file command-line argument. The most accurate alignment will then be output to the file identified by the -out argument. More details of the parameter advising modifications made to Opal are given below.

The advising wrapper, as well as oracle and greedy sets, can be found at facet.cs.arizona.edu.

An updated release of the Opal aligner is available which includes parameter advising performed within the aligner. Opal can now construct alignments under various configuration settings *in parallel* to attempt to come close to producing a parameter-advised alignment in no more wall-time than aligning under a single default parameter.

Because both Opal and Facet are implemented in Java, they can be integrated easily. When an alignment is constructed, a Facet score is automatically generated if secondary structure labeling is given. The secondary structure can be generated using a wrapper for PSIPRED, which will run the secondary-structure prediction and then format the output so it is readable by Opal. For the structure to be able to be used for Facet, you must input this file using the -facet_structure command-line argument. This score is output to standard out. In addition, the score can be printed into the file name by adding the string __FACETSCORE__ to the output file name argument, when the file is created this string is replaced with the computed Facet score.

The new version of Opal also includes the ability to use popular versions of the PAM, BLOSUM and VTML matrices. These can be specified via the -cost command line argument. If the specified cost name is not built in, you can specify a new matrix by giving the file name via the same command line argument. The matrix file should fallow the same formatting convention as BLAST matrices.

If an advisor set of parameter settings is specified using the -advisor_configuration_file command line argument then Opal will construct an alignment for each of the configurations in the file. If in addition a secondary structure prediction is specified, Opal will perform parameter advising. The input advisor set contains a list of parameter settings in 5-tuple format mentioned earlier ($\sigma.\gamma_I.\gamma_T.\lambda_I.\lambda_T$, where σ is the replacement matrix, γ_I and γ_T specify the internal and terminal gap extension penalties and λ_I and λ_T specify the gap open penalties). If advising is performed, the alignment with the highest estimated accuracy is output to the file specified in the -out_best command line argument. In addition, Opal can output the results for each of the configurations specified in the advisor set using -out_config; the filename in that case should contain the string __CONFIG__, which will then be replaced with the parameter setting.

While an alignment must be generated for each parameter setting in the advising set, the construction of these alignments is independent. Because of this we enabled Opal to construct the alignments in the advising set in parallel. Opal will automatically detect how many processors are available and run that many threads to construct alignments, but this can be overridden by specifying a maximum number of threads using the -max_threads command line argument. By doing this, if the number of processors available is larger than the number of parameter choices in the advising set, then the total wall-clock time is close to the time it would take to run the multiple sequence alignment of the input using just a single default parameter choice.

The latest version of the Opal aligner is available at opal.cs.arizona.edu, and the development version of Opal is available on GitHub at git.io/Opal.

Summary

In this chapter, we described the experimental methodology for evaluating the advising accuracy of the Facet estimator, as well as demonstrated the resulting increase in advising accuracy over using a single default parameter choice. The experiments show that Facet is the most accurate estimator for the task of parameter advising. In addition, this chapter demonstrated that estimator-aware greedy advisor sets significantly outperform estimator-oblivious oracle sets.

Chapter 7
Ensemble Multiple Alignment

The multiple sequence alignments computed by an aligner for different *settings* of its parameters, as well as the alignments computed by different *aligners* using their default settings, can differ markedly in accuracy. As we have discussed in previous chapters, *parameter advising* is the task of choosing a parameter setting for an aligner to maximize the accuracy of the resulting alignment. This chapter extends parameter advising to *aligner advising*, which in contrast chooses among a set of aligners to maximize accuracy. In the context of aligner advising, *default* advising selects from a set of aligners that are using their default settings, while *general* advising selects both the aligner and its parameter setting.

In this chapter, aligner advising is used to create a true *ensemble aligner*. Cross-validation experiments on benchmark protein sequence alignments show that parameter advising boosts an aligner's accuracy beyond its default setting for virtually all of the standard aligners currently used in practice. Furthermore, aligner advising with a collection of aligners further improves upon parameter advising with any single aligner, though surprisingly the performance of default advising (choosing only the aligner) on testing data is actually on par with general advising (choosing both an aligner and its parameter settings), due to less overfitting to training data. The performance increase gained using aligner advising is most pronounced on the hardest-to-align sequences.

7.1 Forming the Ensemble Aligner

The preceding chapters have demonstrated how parameter advising can increase the accuracy of multiple sequence alignment. But up to now, the choice of aligner has been static, and the actual impact of parameter advising has thus far only been

Adapted from publication [26].

© Springer International Publishing AG 2017
D. DeBlasio, J. Kececioglu, *Parameter Advising for Multiple Sequence Alignment*,
Computational Biology 26, https://doi.org/10.1007/978-3-319-64918-4_7

85

The AQUA tool of Muller et al. [75] is not a meta-aligner; it uses a method similar to the ensemble aligner described in this chapter, but is limited in that: (1) it only chooses between two aligners using their default parameter settings (namely Muscle and MAFFT), and (2) it uses the NorMD accuracy estimator which, for the task of parameter advising, Chap. 6 showed is dominated by Facet [56].

7.2 Constructing the Universe for Aligner Advising

Chapter 6 shows that parameter advising for a single aligner can dramatically improve alignment accuracy. Using *aligner advising* exploits the strengths of the large number of high quality aligners that have been developed recently by using different aligners to construct our parameter universe, rather than simply alternate parameters for a single aligner. One way to approach aligner advising is to align a set of sequences with several aligners default parameters, then following the procedure presented earlier, choose the computed alignment with the highest predicted accuracy; this is called *default aligner advising*. One potential problem that may arise by simply using the default parameter for each aligner is that the default parameter is intended to work equally well for all input sequences. What makes the parameter advising method so powerful is the ability to have choices that work well on a specific type of input, therefore default aligner advising may suffer from the same problem as using just a single aligner.

The problem of only advising over alignments produced with alignment parameters that work well on average is to examine alignments produced not only with a number of aligners but also consider parameter choices used to produce alignments. The process of advising over aligners and their parameter choices is called *general aligner advising*. To construct a universe of aligners and parameters that work well in concert two basic questions must be answered: (1) Of the many available aligners, which ones might provide the best boost in advising accuracy? and (2) For the aligners chosen how do we create a set of parameter choices for that aligner? Section 7.2.1 deals with the construction of the universe of aligners for default aligner advising, as well as which aligners to use for general aligner advising. Section 7.2.2 then, addresses the question of parameter universe formulation, individually for each aligner.

7.2.1 Determining the Universe of Aligners

For *default aligner advising*, where the advisor set consists of distinct aligners, each using their default parameter setting, an advisor set is learned over a universe containing as many of the commonly-used aligners from the literature as possible. Specifically, the universe for default advising consisted of the following 17 aligners: ClustalW2 [65], Clustal Omega [93], DIALIGN-T [94], DIALIGN-TX [95],

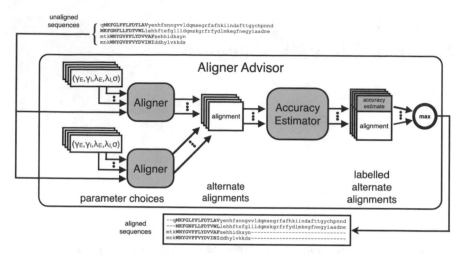

Fig. 7.1 Overview of the ensemble alignment process. An advisor set is a collection of aligners and associated parameter choices. Default aligner advising sets only contain aligners and their default parameter settings, while general aligner advising sets can include non-default parameter settings. An accuracy estimator labels each candidate alignment with an accuracy estimate. The alignment with the highest estimated accuracy is chosen by the advisor.

shown for the Opal aligner. Just as parameter settings can be specialized to perform better on some inputs than others, aligners themselves have input types which are more accurately aligned using their methodology and some for which the aligner produces low quality alignments. By including the choice of aligner in the parameter setting that was defined earlier, we are able to improve the accuracy of an advisor.

In the framework defined in previous chapters a parameter advisor takes a *set* of parameter settings, together with an *estimator* that estimates the accuracy of a computed alignment, and invokes the aligner on each setting, evaluates the accuracy estimator on each resulting alignment, and chooses the setting that gives the alignment of highest estimated accuracy. Analogously, automatically choosing the best aligner for a given input is called *aligner advising*. Figure 7.1 shows an overview of aligner advising. Notice that compared to the Fig. 1.3 the aligner has been moved into the advisor set, and the advisor may use more than one aligner to produce the collection of alternate alignments.

To make this concrete, Fig. 7.2 shows an example of advising on a benchmark set of protein sequences for which a correct reference alignment is known, and hence for which the true accuracy of a computed alignment can be determined. In this example, the Facet estimator is used to estimate the accuracy of two alignments computed by the Opal [107] and MUMMALS [82] aligners. For these

```
d1flma    8   ...   fevlknegvvAIATQgedgphlvntwnsylkv-ldgnrivvpvggmhkteanva-rde  ...   63
d1ci0a   18   ...   tkw-fn--------eakedpret-------------lpeaiTFSS-------Aelpsg    ...   46
d1nrga   16   ...   aaw-fe--------eavqcpdig--------------eanamCLAT-------Ct-rdg   ...   43
d1ejea   22   ...   hriltprptvMVTTVdeegninaapfsftmpvsidppvvafasapdhhtarnie-sth   ...   78
d1i0ra    8   ...   ykisyglyIVTSEsngrkcgqiant---vfqltskpvqiavclnkendthnavk-esg   ...   61
```

(a) Lower-accuracy alignment computed by MUMMALS

```
d1flma    1   ...   ------mlpgtffevlkne-----gvvAIATQg-edgph--lvntwnsylk---vldg   ...   41
d1ci0a   12   ...   d-dpidlftkwfneakedpretlpeaiTFSSAelpsgr----vssrillfk---eldh  ...   59
d1nrga    9   ...   sldpvkqfaawfeeavqcpdigeanamCLATCt-rdgk----psarmlllk---gfgk  ...   56
d1ejea   11   ...   s-mdfedfpvesahriltpr----ptvMVTTVd-eegn----inaapfsftmpvsidp  ...   56
d1i0ra    1   ...   --mdveafykisy-----------glyIVTSE-sngrkcgqiantvfqlt---s-kp   ...   39
```

(b) Higher-accuracy alignment computed by Opal

Fig. 7.2 Aligner choice affects the accuracy of computed alignments. (**a**) Part of an alignment of benchmark sup_125 from the SABRE [102] suite computed by MUMMALS [82] using its default parameter choice; this alignment has accuracy value 28.9%, and Facet estimator value 0.768. (**b**) Alignment of the same benchmark by Opal [106] using its default parameter choice, which has 49.9% accuracy, and higher Facet value 0.781. In both alignments, the positions that correspond to *core blocks* of the reference alignment, which should be aligned in a correct alignment, are highlighted in bold.

two alignments, the one of higher Facet value has higher true accuracy as well, so an advisor armed with the Facet estimator would in fact output the more accurate alignment to a user.

For a collection of aligners, this kind of advising is akin to an *ensemble* approach to alignment, which selects a solution from those output by different methods to obtain in effect a new method that ideally is better than any individual method. Ensemble methods have been studied in machine learning [115], which combine the results of different classifiers to produce a single output classification. Typically such ensemble methods from machine learning select a result by *voting*. In contrast, an advisor combines the results of aligners by selecting one via an *estimator*.

This chapter extends the framework of parameter advising to aligner advising, and obtains by this natural approach a true *ensemble aligner*. Moreover as our experimental results show, the resulting ensemble aligner is significantly more accurate than any individual aligner.

Recall from Sect. 1.3.3 that there have been several so-called *meta-alignment* techniques in use, these methods *combine* the output of several aligners to produce a single alignment. The two most popular are M-Coffee from Wallace et al. [103], and MergeAlign from Collingridge and Kelly [19].

One problem that both M-Coffee and MergeAlign face is that for hard to align sets of sequences, few aligners will produce a high accuracy alignment. By considering all of the inputs to be equally accurate, they drown out the signal of the one aligner that did align the sequences correctly. As Sect. 7.3.3 later shows, when run on the same set of aligners, both of these approaches are strongly dominated by the ensemble approach of this chapter.

FSA [12], Kalign [66], MAFFT [55], MUMMALS [82], Muscle [37], MSAProbs [69], Opal [106], POA [68], PRANK [71], PROBALIGN [89], ProbCons [34], SATé [70], and T-Coffee [78].

7.2.2 Determining the Universe of Parameter Settings

For *general aligner advising*, a parameter universe was created for each of the aligners used for default aligner advising; the full list of parameters is found in Table 7.1. For each aligner, parameter settings were enumerated by forming a cross product of values for each of its tunable parameters. The values for each tunable parameter were determined by one of two ways. For aligners with available web-servers or GUIs with predefined options (namely Clustal Omega, ProbCons, and SATé), all values recommended for each parameter were used (or as many as possible when this list was too long). For all other aligners, either one or two values above and below the default value for each parameter were chosen, to attain a cross product with less than 200 parameter settings. If a range was specified for a numeric parameter, values were chosen to cover this range as evenly as possible. For non-numeric parameters, all available options were used.

7.3 Evaluating Ensemble Alignment

The performance of advising is evaluated through experiments on the same collection of protein multiple sequence alignment benchmarks described in Chap. 6. The experiments in this section compare the accuracy of parameter and aligner advising to the accuracy of individual aligners using their default parameter settings.

Recall that the collection of benchmarks consists of 861 reference alignments. Also as mentioned earlier it is common in benchmark suites, easy-to-align benchmarks are highly over-represented in this collection, compared to hard-to-align benchmarks. Earlier the bias was corrected by binning benchmarks based on the accuracy of the alignment produced using the Opal aligner's default parameters. Since more than one aligner is now being considered a more general method for bias correction is used.

To correct for this bias when evaluating average advising accuracy, the 861 benchmarks in the collection were again binned by *difficulty*, where the difficulty of a benchmark is now defined as the average accuracy under the most commonly-used aligners, namely Clustal Omega, MAFFT, and ProbCons, using their default parameter settings. As in the previous chapter the full range [0, 1] of accuracies was then divided into 10 bins with difficulties $[(j-1)/10, j/10]$ for $j = 1, \ldots, 10$. The weight w_i of benchmark B_i falling in bin j used for training is $w_i = (1/10)(1/n_j)$, where n_j is the number of benchmarks in bin j. These weights w_i are such that each difficulty bin contributes equally to the advising

Table 7.1 Universe of parameter settings for general aligner advising

Aligner	Parameter settings	Tunable parameters	Version	Parameter name	Default value, v	Alternate values
Clustalw2 [65]	162	5	2.1	Substitution matrix	GONNET	PAM, BLSM
				Gap open penalty	10	5, 20
				Gap extension penalty	0.2	0.1,0,4
				Gap distance	4	2,8
				End gaps	off	on
Clustal Omega [93]	120[a]	5	1.2.0	Number of guide tree iterations	0	1, 3, 5
				Number of HMM iterations	0	1, 3, 5
				Number of combined iterations	0	1, 3, 5
				Distance matrix calculations, initial	mBed	Full alignments
				Distance matrix calculations, iterations	mBed	Full alignments
DIALIGN-T [94]	162	5	0.2.2	Substitution matrix	BLSM62	BLSM75, BLSM90
				Overlap	0	1
				Global minimum	40	20,60
				Threshold	4	2,6
				Even threshold	4	2,6
DIALIGN-TX [95]	162	5	1.0.2	Substitution matrix	BLSM62	BLSM75, BLSM90
				Probability distribution	BLOSUM.diag_prob_t10	BLOSUM75.diag_prob_t2
				Sensitivity	0	1,2
				Threshold	4	2,6
				Even threshold	4	2,6

Aligner	Parameter settings	Tunable parameters	Version	Parameter name	Default value, v	Alternate values
FSA [12]	200	5	1.15.3	Estimate indel probabilities	on	off
				Estimate emission probabilities	on	off
				Regularize learned emission and gap probabilities	on	off
				Maximum number of iterations of EM	3	1,5,10,20
				Minimum fractional increase per iteration	0.1	0.01,0.05,0.15,0.2
Kalign [66]	162	4	2.04	Gap open penalty	55	40, 70
				Gap extension penalty	8.5	7, 10
				Terminal gap penalty	4.25	3.5, 5
				Bonus	no	yes
MAFFT [55]	175	3	6.923b	Substitution matrix	BLSM62	BLSM30, BLSM45, BLSM80, VTML120, VTML200, VTML350
				Gap open penalty	1.53	$\frac{1}{4}v, \frac{1}{2}v, \frac{3}{2}v, 2v$
				Gap extension penalty	0.123	$\frac{1}{2}v, 2v, 4v$
MSAProbs [69]	175	3	0.9.7	Passes of consistency transformation	2	0,1,3,4,5
				Passes of iterative-refinement	10	0,2,5,20,50,75,100
Muscle [37]	160	3	3.8.31	Profile score	Log-expectation: VTML240	Sum-of-pairs: PAM200, VTML240
				Objective function[b]	spm	dp, ps, sp, spf, xp
				Gap open penalty, profile dependent	$\gamma = v^c$	$\frac{1}{2}v, \frac{3}{4}v, \frac{5}{4}v, \frac{3}{2}v$
				Gap extension penalty	γ	$\frac{1}{2}\gamma$

(continued)

Table 7.1 (continued)

Aligner	Parameter settings	Tunable parameters	Version	Parameter name	Default value, v	Alternate values
MUMMALS [82]	29	3[d]	1.01	Differentiate match states in unaligned regions	yes	no
				Solvent accessibility categories	1	2, 3
				Secondary structure types	3	1
Opal [106]	162	5	3.0b	Substitution matrix	VTML200[e]	BLSM62[e], VTML40[e]
				Internal gap open penalty	$\gamma = 45$	70, 95
				Terminal gap open penalty	0.4γ	0.05γ, 0.75γ
				Internal gap extension penalty	$\lambda = 42$	40, 45
				Terminal gap extension penalty	$\lambda - 3$	λ
POA [68]	144	2	1.0.0	Substitution matrix	BLSM80	BLSM62, VTML120, VTML200
				Gap penalty 1	12	3, 24
				Gap penalty 2	6	1, 3
				Progressive alignment	yes	no
				Global alignment	yes	no
PRANK [71]	165	3	0.140603	Gap rate	0.005	$\frac{1}{5}v, \frac{1}{2}v, \frac{3}{2}v, 2v$
				Gap extension	0.5	$\frac{1}{5}v, \frac{1}{2}v, \frac{3}{2}v, 2v$
				Terminal gaps	Alternate scoring	Normal scoring
				Force insertions to be always skipped	yes	no
				Iterations	5	1
ProbCons [34]	168[f]	3	1.4	Consistency repetitions	2	0, 1, 3, 4, 5
				Iterative refinement repetitions	100	0, 500, 1000
				Pre-training repetitions	0	1, 2, 3, 4, 5, 20

Aligner	Parameter settings	Tunable parameters	Version	Parameter name	Default value, v	Alternate values
PROBALIGN [89]	124	3	1.12	Thermodynamic temperature	5	1, 3, 5, 10
				Gap open	22	1, 11, 33, 55
				Gap extension	1	0.25, 0.5, 1.5, 3
SATé [70]	190	4	2.2.7	Aligner	MAFFT	Opal, PRANK, ClustalW
				Merger	Muscle	Opal
				Tree generator	FastTree	RAxML
				Tree model	G20/Gamma[g]	CAT
				Tree matrix	WAG	JTT
				Mode[h]	fast	ML, simple
T-Coffee [78]	180	3	10.00.r1613	Substitution matrix	BLSM62	BLSM40, BLSM80, PAM120, PAM250, PAM350
				Gap open	0	$-50, -500, -1000, -5000$
				Gap extension	0	$-1, -3, -5, -7, -10$
Total	2529					

[a] Parameter settings retrieved from the Clustal Omega web-server at EBI (www.ebi.ac.uk/Tools/msa/clustalo)

[b] sp: sum-of-pairs score; spf: dimer approximation of sum-of-pairs score; spm: input dependent (sp if input is less than 100 sequences, spf otherwise); dp: dynamic programming score; ps: average profile sequence score; xp: cross profile score

[c] Default values for the gap open penalty are -2.9 when the log-expectation profile is chosen, -1439 for sum-of-pairs using PAM200, and -300 for sum-of-pairs using VTML240. Alternate values are multiples of this default value

[d] MUMMALS is distributed with 29 precomputed hidden Markov models, each of which is associated with a setting of three tunable parameters

[e] The substitution matrices used by Opal are shifted, scaled, and rounded to integer values in the range [0, 100]

[f] Parameter settings retrieved from the ProbCons web-server at Stanford (probcons.stanford.edu)

[g] G20 is the default for FastTree while Gamma is the default for RAxML

[h] Runtime settings retrieved from the SATé GUI. Fast: stop after one iteration of non-improvement, return the best alignment seen; ML: run ten iterations, return best alignment seen; simple: stop after ten iterations of non-improvement, return final alignment

objective function $f(P)$. Note that with this weighting, an aligner that on every benchmark gets an accuracy equal to its difficulty, will achieve an average advising accuracy of roughly 50%.

7.3.1 Parameter Advising

This section presents results of parameter advising for a single aligner using the Facet estimator. The coefficients for Facet were learned by difference fitting on computed alignments obtained using the oracle set of cardinality $k = 50$ found for the parameter universe for each aligner (except MUMMALS where $k = |U| = 29$). The threshold-difference pair parameters used were $\epsilon = 0.0015$ and $\ell = 10,000$; all of these meta-parameter choices were found by first enumerating all reasonable values for each meta-parameter and choosing combination of settings that had the highest training accuracy averaged across the four most accurate aligners (MUMMALS, Opal, PROBALIGN, and SATé) and the general aligner advisor. Given this estimator, greedy advisor sets were constructed for each aligner.

Figure 7.3 shows the accuracy of parameter advising using greedy advisor sets of cardinality $k \leq 25$, for each of the ten aligners in Table 7.1, under 12-fold cross-validation. The plot shows advising accuracy on the testing data, averaged over all benchmarks and folds.

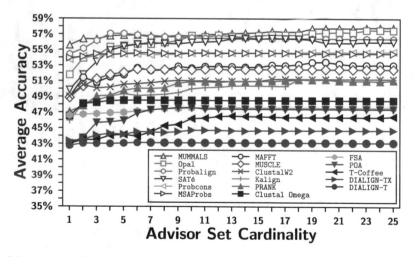

Fig. 7.3 Accuracy of parameter advising using Facet. The plot shows advising accuracy for each aligner from Table 7.1, using parameter advising on greedy sets with the Facet estimator learned by difference fitting. The horizontal axis is the cardinality of the advisor set, and the vertical axis is the advising accuracy on testing data averaged over all benchmarks and folds, under 12-fold cross-validation.

Almost all aligners benefit from parameter advising, though their advising accuracy eventually reaches a plateau. While our prior chapters showed that parameter advising boosts the accuracy of the Opal aligner, Fig. 7.3 shows this result is not aligner dependent.

7.3.2   Aligner advising

To evaluate aligner advising, a similar approach was used; constructing an oracle set of cardinality $k = 17$ for default aligner advising, and $k = 50$ for general, over the union of the universe for default advising from Sect. 7.2.1, and the universe for general advising from Sect. 7.2.2; learning coefficients for Facet using difference fitting; and constructing greedy sets using Facet for default and general advising.

Figure 7.4 shows the accuracy of default and general advising using greedy sets of cardinality $k \leq 25$, along with the four best parameter advising curves from Fig. 7.3, for MUMMALS, Opal, PROBALIGN, and SATé. The plot shows advising accuracy on testing data, averaged over benchmarks and folds.

Default advising actually has higher accuracy than general advising at low cardinalities (as later sections will show this is most likely due to overfitting to the training benchmarks), but at cardinalities $k \geq 9$ general advising provides a boost in accuracy. Both methods have higher advising accuracies than using only a

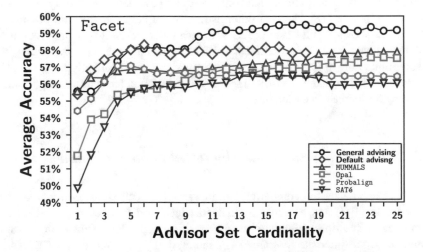

Fig. 7.4 Aligner advising and parameter advising using Facet. The plot shows default and general aligner advising accuracy, and parameter advising accuracy for MUMMALS, Opal, PROBALIGN, and SATé, using the Facet estimator. The horizontal axis is the cardinality of the advisor set, and the vertical axis is advising accuracy on testing data, averaged over all benchmarks and folds under 12-fold cross-validation.

single aligner across the range of cardinalities. This is most pronounced at higher cardinalities where general aligner advising sees a boost in accuracy of over 2% (for $11 \leq k \leq 18$).

7.3.2.1   Testing the Significance of Improvement

A one-tailed Wilcoxon sign test [109] was used to test the statistical significance of the improvement in general advising accuracy over using a single default aligner. Performing this test in each difficulty bin yielded a significant improvement in accuracy ($p < 0.05$) on benchmarks with difficulty $[0, 0.1]$ at all cardinalities $5 \leq k \leq 25$ ($p < 0.01$ for $k \geq 6$), on benchmarks with difficulty in the range $(0.5, 0.6]$ for cardinalities $8 \leq k \leq 22$, and in the range $(0.3, 0.4]$ for $14 \leq k \leq 25$.

The significance of the improvement of default advising over the best parameter advisor at each cardinality k (namely MUMMALS) was also tested; this revealed that at cardinality $k \geq 6$ there is again significant improvement ($p < 0.05$) on benchmarks with difficulty $[0, 0.1]$ ($p < 0.01$ when $k \geq 11$). General advising also shows significant improvement over parameter advising on at least one other bin as well at cardinality $k \geq 6$.

7.3.2.2   Advising with an Alternate Estimator

Parameter advising and aligner advising were also evaluated on greedy sets using the TCS estimator [18] (the best other estimator for advising from the literature). Figure 7.5 shows results using TCS for parameter advising (on the four most accurate aligners), and for general and default aligner advising. Note that while TCS is sometimes able to increase accuracy above using a single default parameter, this increase is smaller than for Facet; moreover, TCS often has a decreasing trend in accuracy for increasing cardinality.

7.3.3   Comparing Ensemble Alignment to Meta-alignment

Another approach to combining aligners is so-called *meta-alignment* of M-Coffee [103] and MergeAlign [19] (described in Sect. 7.1). M-Coffee computes a multiple alignment using position-dependent substitution scores obtained from alternate alignments generated by a collection of aligners. MergeAlign computes alignment as the maximum weight path in the combined partial order alignment graph. By default, M-Coffee uses the following eight aligners: Clustal2, T-Coffee, POA, Muscle, MAFFT, Dialign-T, PCMA [84], and ProbCons. The tool also allows use of Clustal, Clustal Omega, Kalign, AMAP [91], and Dialign-TX. Figure 7.6 shows the average accuracy of M-Coffee, MergeAlign, and our ensemble approach

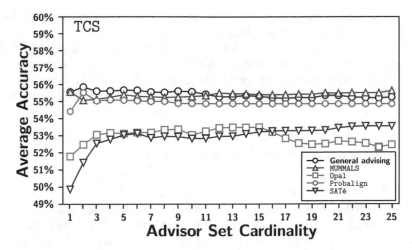

Fig. 7.5 Aligner advising and parameter advising using TCS. The plot shows default and general aligner advising accuracy, and parameter advising accuracy for Opal, MUMMALS, PROBALIGN, and ProbCons, using the TCS estimator. The horizontal axis is the cardinality of the advisor set, and the vertical axis is advising accuracy on testing data averaged over all benchmarks and folds under 12-fold cross-validation.

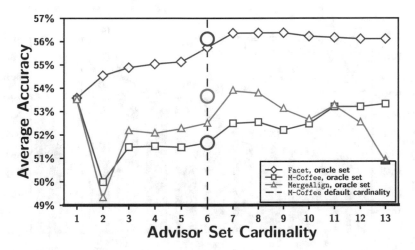

Fig. 7.6 Accuracy of aligner advising compared to M-Coffee and MergeAlign. The plot shows average accuracy for aligner advising using Facet, and meta-alignment using M-Coffee and MergeAlign, on oracle sets of aligners. Performance on the default M-Coffee and MergeAlign set of eight aligners is indicated by large circles on the dotted vertical line. The horizontal axis is cardinality of the oracle sets, and the vertical axis is average accuracy on testing data over all benchmarks and folds under 12-fold cross-validation.

with Facet, using the default aligner set of M-Coffee (the dotted vertical line with large circles), as well as oracle sets constructed over this M-Coffee universe of 13 aligners. Notice that at all cardinalities our ensemble aligner substantially outperforms meta-alignment even on the subset of aligners recommended by the M-Coffee developers. Also note that greedy sets cannot be constructed for M-Coffee because it does not have an explicit score used to select alignments.

7.3.4   Advising Accuracy Within Difficulty Bins

Figure 7.7 shows advising accuracy within difficulty bins for general aligner advising compared to using the default parameter settings for the three aligners with highest average accuracy, namely MUMMALS, Opal, and PROBALIGN. The figure displays the default advising result from Sect. 7.3.2 at cardinality $k = 15$. The bars in the chart show average accuracy over the benchmarks in each difficulty bin, as well as the average accuracy across all bins. (The number of benchmarks in each bin is in parentheses above the bars). Note that aligner advising gives the greatest boost for the hardest-to-align benchmarks: advising provides an increase of at least 2.5% on the bottom six bins over the best aligner using its default parameter setting.

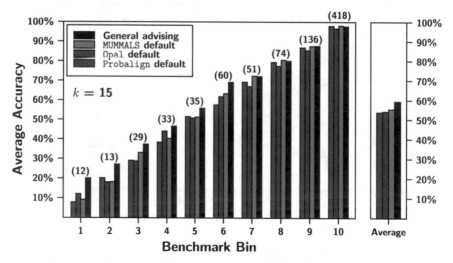

Fig. 7.7 Accuracy of aligner advising, and aligners with their default settings, within difficulty bins. In the bar chart on the left, the horizontal axis shows all ten benchmark bins, and the vertical bars show accuracy averaged over just the benchmarks in each bin. The accuracy of general advising using the Facet estimator is shown for the greedy sets of cardinality $k = 15$, along with the accuracy of the default settings for PROBALIGN, Opal, and MUMMALS. The bar chart on the right shows accuracy uniformly averaged over the bins. In parentheses above the bars are the number of benchmarks in each bin.

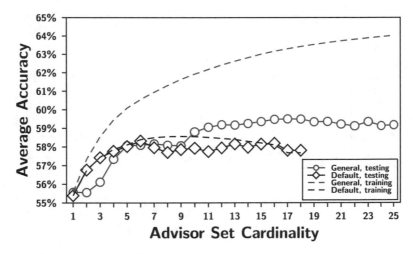

Fig. 7.8 General and default aligner advising on training and testing data. The plot shows general and default aligner advising accuracy using Facet. Accuracy on the training data is shown with dashed lines, and on the testing data with solid lines. The horizontal axis is cardinality of the advisor set, and the vertical axis is advising accuracy averaged over all benchmarks and folds under 12-fold cross-validation.

This is particularly pronounced in the bottom two bins where there is an almost 8% and 7% increase in accuracy respectively.

7.3.5   Generalization of Aligner Advising

The results thus far have shown advising accuracy averaged over the testing data associated with each fold. This section compares the training and testing advising accuracy to assess how our method might generalize to data not in our benchmark set.

Figure 7.8 shows the average accuracy of default and general aligner advising on both training and testing data. There is a drop in training accuracy for default advising with increasing cardinality, though after its peak an advisor using greedy sets should remain flat in training accuracy as cardinality increases, when using a strong estimator. This drop in training accuracy is due to the fact that the advisor is forced to have a cardinality k *equal* to some value, rather than being able to choose a smaller cardinality that has higher accuracy. Additionally, note that the drop between training and testing accuracy is much larger for general advising than for default advising, resulting in general advising performing only as good as default advising, though its training accuracy is much higher. This indicates that general advising is strongly overfitting to the training data, but could potentially achieve much higher testing accuracy.

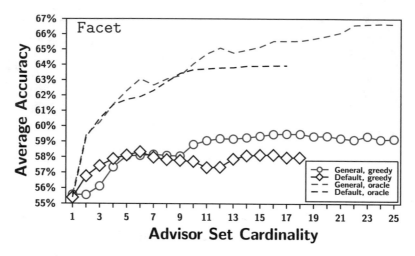

Fig. 7.9 Accuracy of aligner advising using a perfect estimator. The plot shows advising accuracy for default and general aligner advising, both on oracle sets for a perfect estimator, and on greedy sets for the Facet estimator. The horizontal axis is the cardinality of the advisor set, and the vertical axis is advising accuracy on testing data averaged over all benchmarks and folds under 12-fold cross-validation.

7.3.6 Theoretical Limit on Advising Accuracy

An *oracle* is an advisor that uses a perfect estimator, always choosing the alignment from a set that has highest true accuracy. To examine the theoretical limit on how well aligner advising can perform, the accuracy of aligner advising using Facet is compared with the performance of an oracle. Figure 7.9 shows the accuracy of both default and general aligner advising using greedy sets, as well as the performance of an oracle using oracle sets computed on the default and general advising universes. (Recall an *oracle set* is an optimal advisor set for an oracle.) The plot shows advising accuracy on testing data, averaged over all benchmarks and folds. The large gap in performance between the oracle and an advisor using Facet shows the increase in accuracy that could potentially be achieved by developing an improved estimator.

7.3.7 Composition of Advisor Sets

Table 7.2 lists the greedy advisor sets for both default and general advising for all cardinalities $k \leq 10$. A consequence of the construction of greedy advisor sets is that the greedy set of cardinality k consists of the entries in a column in the first k rows of the table. The table shows these sets for just one fold from the 12-fold cross-validation. For general advising sets, an entry specifies the aligner that is used, and for aligners from the general advising universe, a tuple of parameter values in

Table 7.2 Greedy default and general advising sets

	Default advising	General advising
1	MUMMALS	MUMMALS (0.2, 0.4, 0.6, 1, 2, 3)
2	Opal	Opal (VTML200, 45, 2, 45, 45)
3	PROBALIGN	ProbCons (2, 100, 0)
4	ClustalW2	MUMMALS (0.15, 0.2, 0.6, 1, 1, 1)
5	Kalign	PROBALIGN (10, 33, 3)
6	MSAProbs	Opal (VTML40, 95, 4, 42, 39)
7	Muscle	PROBALIGN (7, 33, 0.25)
8	Clustal Omega	Kalign (55, 8.5, 4.25, no)
9	FSA	SATé (Opal, Opal, RAxML, Gamma, WAG, fast)
10	T-Coffee	ClustalW2 (BLSM, 10, 0.1, 8, on)

the order listed in Table 7.1. The one exception is MUMMALS, whose 6-tuple comes from its predefined settings file, and whose last three elements correspond to the three parameters listed in Table 7.1. It is interesting that other than ProbCons and Kalign, the general advising set does not contain any aligner's default parameter settings, though many of the chosen values are close to the default setting.

7.3.8 Running Time for Advising

The time to evaluate the Facet estimator on an alignment was compared to the time needed to compute that alignment by the three aligners used for determining alignment difficulty: Clustal Omega, MAFFT, and ProbCons. To compute the average running time for these aligners on a benchmark, the total time for each of these aligners to align all 861 benchmarks was measured on a desktop computer with a 2.4 GHz Intel i7 8-core processor and 8 Gb of RAM. The average running time for Clustal Omega, MAFFT, and ProbCons was less than 1 second per benchmark, as was the average running time for Facet. As stated in Chap. 3 the time complexity for Facet is dependent on the number of columns in an alignment, and should take relatively less time than computing an alignment for benchmarks with long sequences; the standard benchmark suites tend to include short sequences, however, which are fast to align. This time to evaluate Facet does not include the time to predict protein secondary structure, which is done once for the sequences in a benchmark, and was performed using PSIPRED [51] version 3.2 with its standard settings. Secondary structure prediction with a tool like PSIPRED has a considerably longer running time than alignment, due to an internal PSI-BLAST search during prediction; on average, PSIPRED took just under 6 min per benchmark to predict secondary structure.

7.4   Software

The new ensemble aligner is implemented using `Perl` as a wrapper around the various underlying aligners. The `Perl` programs can be used in two ways:

(1) ***Using a predefined set of programs***   The `Facet` release comes with two applications, `default_ensemble_alignment.pl` and `ensemble_alignment.pl`, which can be used to run default and general ensemble alignment on sets learned from all training benchmarks. To use these applications, you provide the set cardinality you would like to use, the unaligned sequences, and predicted secondary structure. The application then runs each program in order, and outputs the result to standard out.

(2) ***Using a program that accepts an advisor set***   We have also included an application `ensemble_alignment_from_set.pl` that accepts the input unaligned sequences, secondary structure prediction, and an advisor set similar to the one defined in earlier chapters for parameter advising. The advisor set contains the aligner and parameter setting information that will be used to run the aligners. Each line of the file contains one aligner and parameter setting in the format A_S where A is the aligner name and S is the tuple of parameter settings for that aligner separated by "." characters; for example, the default `Opal` parameter setting would be `Opal_VTML200.45.11.42.41`. Parameter advising sets for each of the applications tested, as well as default and general advising parameter sets in the proper format, can be found on the `Facet` website.

For both of these methods, the applications must be edited if any of the applications being used are not in the default installation location.

Summary

This chapter has described how to extend parameter advising to *aligner advising*, to yield a true *ensemble aligner*. Parameter advising gives a substantial boost in accuracy for nearly all aligners currently used in practice. Furthermore, default and general aligner advising both yield further boosts in accuracy, with default advising having better generalization. As these results indicate, ensemble alignment by aligner advising is a promising approach for exploiting future advances in aligner technology.

Chapter 8
Adaptive Local Realignment

Mutation rates differ across the length of most proteins, but when multiple sequence alignments are constructed for protein sequences, a single alignment parameter choice is often used across the entire length. This chapter reviews an approach called *adaptive local realignment*, which is the first method to compute protein multiple sequence alignments using diverse parameter settings for different regions of the input sequences. In this way, parameter choices can vary across the length of a protein to more closely model the local mutation rate. Using adaptive local realignment boosts alignment accuracy over using a default parameter choice. In addition, when adaptive local realignment is combined with global parameter advising, we see an increase in accuracy of almost 24% over the default parameter choice on hard-to-align benchmarks.

8.1 Relating Protein Mutation Rates to Alignment

Since the 1960s, it has been known that proteins can have distinct mutation rates at different locations along the molecule [43]. The amino acids at some positions in a protein may stay unmutated for long periods of time, while other regions change a great deal (often called "hypermutable" regions). This has led to methods in phylogeny construction that take variable mutation rates into account when building trees from sequences [113]. In multiple sequence alignment, however, to our knowledge variation in mutation rates across sequences has yet to be exploited to improve alignment accuracy. Multiple sequence alignments are typically computed using a single setting of values for the parameters of the alignment scoring function. This single parameter setting affects how residues across a protein are aligned, and implicitly assumes a uniform mutation rate. In contrast, the approach discussed in

---

Adapted from publication [30].

© Springer International Publishing AG 2017   103
D. DeBlasio, J. Kececioglu, *Parameter Advising for Multiple Sequence Alignment*,
Computational Biology 26, https://doi.org/10.1007/978-3-319-64918-4_8

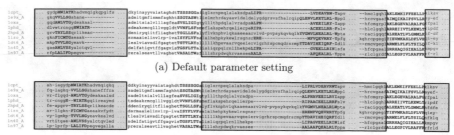

(a) Default parameter setting

(b) Using adaptive local realignment

Fig. 8.1 Impact of adaptive local realignment. The figure shows portions of an alignment of benchmark BB11007 from the BAliBASE suite, where the highlighted amino acids in red uppercase are from the core columns of the reference alignment, which should be aligned in a correct alignment. (**a**) The alignment computed by Opal using its optimal default parameter setting (VTML200, 45, 11, 42, 40) across the sequences, with an accuracy of 89.6%. The regions of the alignment in gray boxes are automatically selected for realignment. (**b**) The outcome of using adaptive local realignment, with an improved accuracy of 99.6%. The realignments of the three regions use alternate parameter settings (BLOSUM62, 45, 2, 45, 42), (BLOSUM62, 95, 38, 40, 40), and (VTML200, 45, 18, 45, 45), respectively, which increase the accuracy of these regions.

this chapter identifies alignment regions that may be misaligned under a single parameter setting, and finds alternate parameter settings that may more closely match the local mutation rate of the sequences.

This chapter describes a method that takes a given alignment and attempts to improve its overall accuracy by replacing sections of it with better subalignments, as demonstrated in Fig. 8.1. The top alignment of the figure was computed using a single parameter setting: the optimal default setting of the Opal aligner [106]. The bottom alignment is obtained by adaptive local realignment, taking the top alignment, automatically identifying the sections in gray boxes, and realigning them using alternate parameter settings, as described later in Sect. 8.2. This increases the overall alignment accuracy by 10%, as most of the misaligned core blocks (highlighted in red uppercase) are now corrected.

Recall from Sect. 1.3.5 that there are two main realignment methods:

(a) *horizontal realignment* removes columns of an alignment, then realigns them, and

(b) *vertical realignment* realigns groups of whole sequences.

Conceptually, realignment attempts to *correct* errors in existing alignments that were made during the alignment process. Several tools attempt to *avoid* making these errors in the first place, by adjusting parameter values along the sequences during alignment construction. For example, PRANK [71] uses a multi-level HMM that effectively chooses the alignment scoring function at each position. T-Coffee [78] uses consistency between pairwise alignments to create position-specific substitution scores. In fact, even the early tool ClustalW [98] adjusted positional gap-penalties based on pairwise sequence characteristics. Nevertheless, these tools which adjust positional alignment scores all tend to attain lower accuracies on protein benchmarks than the aligners which make no positional adjustments that we compare against in our experiments.

By comparison, adaptive local realignment is a vertical approach that: aims to improve alignment accuracy, applies to any alignment method that has tunable parameters, and to our knowledge is the first approach to alignment that can automatically adapt to varying mutation rates along a protein.

8.2   Adaptive Local Realignment

The *adaptive local realignment* method was developed to overcome the issue of protein sequences being non-homogeneous and having regions that may require different alignment parameters choices. Adaptive local realignment uses some of the same basic principles that have been shown to work well for global parameter advising. The techniques described in previous chapters are applied locally to choose the best alignment parameters for a subset of columns of an alignment.

The adaptive local realignment method for an alignment can be broken down into two steps: (1) choosing regions of the alignment that are correctly aligned which we should save, and (2) producing a new alignment for those regions that are not correctly aligned.

Similar to global parameter advising, local realignment relies on a set of alternate parameter choices and an accuracy estimator.

8.2.1   Identifying Local Realignment Regions

Just as with global alignments, a known reference alignment is not available, so identifying the alignment columns that are recovered correctly in a computed alignment is not possible. Therefore an accuracy estimator E is needed to identify the regions of a given alignment that are going to be retained (and those that will be realigned). The estimated accuracy is calculated for sliding windows that span the length of the alignment (see Fig. 8.2a). The window size is defined as a fraction $w \leq 1$ of the total length of the alignment. The window size w must be chosen carefully because the accuracy estimator has features that are global calculations of an alignment. A larger sliding window will provide more context at each position and should provide a better estimate of accuracy. At the same time, if the window is too large there will not be enough granularity to identify the transitions between correct and incorrectly aligned columns. To account for very short and very long alignments a minimum w_{min} and maximum w_{max} window size is specified.

The previous step yields scores for approximately $\frac{1}{w}$ windows that overlap each column of an alignment. A score for each column is calculated as a sum of these

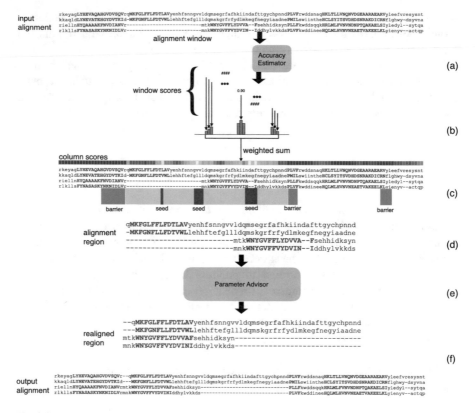

Fig. 8.2 The adaptive local realignment process. (**a**) Calculate a Facet score for a sliding window across at the input alignment. (**b**) A score for each column from the set of window scores is calculated using a weighted sum of the values for all windows that overlap that column. (**c**) Columns that have a column score value greater than τ_G are labeled as barriers and then columns with value less than τ_B are used as seeds for realignment regions. (**d**) These seeds are then extended in both directions until they reach a barrier column to define a realignment region that is extracted from the alignment. (**e**) The unaligned subsequences defined by this region are then realigned using a parameter advisor. (**f**) Once the most accurate realignment of the region is found it is reinserted into the input alignment replacing the section that was removed.

window scores, weighted according to the distance to the center column of that window (see Fig. 8.2b). The weight of the contribution that overlaps a column is determined using a geometric distribution, with a decay factor $d \leq 1$ centered on the middle column of the window. As d approaches 1, a column gets equal weight from all covering windows; as d approaches 0, the score is dependent only on the window centered at that column.

Two thresholds τ_B and τ_S on the column scores are then calculated based on: the percentage of columns from the original alignment we would like to keep, T_B;

and the percentage of columns we will use to seed realignment regions, T_S. The thresholds are set so that at least $\lceil \ell \ T_B \rceil$ columns have score that are above τ_B, and at least $\lceil \ell \ T_S \rceil$ columns have scores below τ_S. All columns with scores at least τ_B are labeled "barriers" and all columns with scores at most τ_S are labeled "seeds" (see Fig. 8.2b).

The alignment regions that will be realigned are defined in a manner similar to the *seed and extend* model used in search programs like BLAST. Each seed is the start of such an alignment region. This region is then extended to include any other seed or unlabeled column to the left and right. This expansion continues until a barrier column (or the end of the alignment) is reached in both directions.

The barrier columns will never be included in an alignment region that will be realigned. In this way at least T_B percent of the columns from the original alignment are guaranteed to remain, and there will always be at least one region to realign.

8.2.2 Local Parameter Advising on a Region

During local advising a new alignment is constructed that contains all of the columns surrounded by only barrier columns, and more accurate alignments of the columns that are covered by alignment regions (if a more accurate alignment can be found).

For each alignment region the sub-alignment in the contained columns is extracted and its Facet score is calculated (see Fig. 8.2d). Later other alternate alignments will be compared with this base accuracy. The substrings comprising this region are then collected. This set of unaligned sequences becomes the input to parameter advising.

The parameter advising method described in Sect. 1.2 and Fig. 1.3 is used to find new alignments with one exception. As noted in Chap. 6 the Opal aligner has 5 tunable parameters: the replacement matrix as well as two internal and two terminal gap costs. For those regions that do include the alignment terminals (the first or last column of the input alignment), the terminal gap penalties are replaced with the corresponding internal gap cost. For those regions that do include terminals the terminal gap penalty is used only on the one side that is terminal in the context of the global alignment. Note that an alignment region as defined will never include both terminals.

The advisor's choice is then compared with the original alignment of this region; if the accuracy of the new alignment is higher, the columns covered by the alignment region are removed from the input alignment and replaced with the new alignment of this region (see Fig. 8.2f).

As a final step the accuracy of the new alignment is compared to the input alignment. The more accurate global alignment is returned.

8.2.3 Iterative Local Realignment

Once a new global alignment of the input sequences is produced, adaptive local realignment can be repeated to continue to refine the alignment. Using the same methods described earlier, the Facet score is computed on windows of this new alignment, to get column scores, and alignment regions are defined for parameter advising to realign. This process iterates until a user defined maximum number of iterations is reached. Note that the local advising procedure may reach a point where none of the misaligned regions are replaced; even though continuing to iterate will not effect the output, iteration is halted to reduce the running time.

8.2.4 Combining Local with Global Advising

Local advising is a method for improving the accuracy of an existing alignment. Recall from previous chapters that using parameter choices other than the default can greatly increase the alignment accuracy for some inputs. Therefore global advising can be used to find a more accurate starting point for adaptive local realignment.

Local and global advising can be combined in two ways.

(1) **Local advising on *all* global alignments:** using adaptive local realignment on each of the alternate alignments produced within global parameter advising, then choosing among all $2|P|$ alternate alignments ($|P|$ unaltered global alignments and $|P|$ locally advised alignments), and

(2) **Local advising on *best* global alignment:** choosing the best global alignment, then using adaptive local realignment to boost its accuracy.

8.3 Assessing Local Realignment

The performance of adaptive local realignment is evaluated through experiments on the same collection of protein benchmarks described in Chap. 6. The experiments in this section compare the accuracy of alignments produced using Opal's default parameter settings with those that have been produced using local and global advising, both individually and in combination.

Recall that the collection of benchmarks consists of 861 reference alignments. Also as mentioned earlier, as is common in benchmark suites, easy-to-align benchmarks are highly over-represented in this collection, compared to hard-to-align benchmarks. The bias is corrected by binning benchmarks based on the accuracy of the alignment produced using the Opal aligner's default parameters, as it was in Chap. 6.

Table 8.1 Tunable parameters for adaptive local realignment

Tunable parameter	Chosen value	Tested values
Estimator window size percentage, w	30%	10%, 20%, 30%, ..., 90%
Minimum window sizes, w_{min}	10	5, 10, 20, 30
Minimum window sizes, w_{max}	30	30, 50, 75, 100, 125
Good column label percentages, B_G	10%	5%, 10%, 20%, 30%, ..., 70%
Bad column label percentages, B_B	30%	5%, 10%, 20%, 30%, ..., 70%
Gamma decay value, d	0.9	0.5, 0.66, 0.9, 0.99

We trained the estimator coefficients for Facet using the difference fitting method described in Chap. 2 on the training sets described above. We found that there was very little change in coefficients between the training folds so for ease of experimentation we use the estimator coefficients that are released with the newest version of Facet, which were trained on all available benchmarks.

Table 8.1 shows the six tunable parameters for adaptive local realignment as described in the previous section. A selection of values was chosen for each of these parameters and each combination of variables was used to apply local advising to the default alignments from Opal. In the experiments shown below the chosen settings for each of the tunable parameters (shown in the center column of the table) are the settings that gave the highest accuracy on the *training* benchmarks described above.

8.3.1 Effect of Local Realignment Across Difficulty Bins

Figure 8.3 shows the alignment accuracy across difficulty bins for default alignments from Opal, local advising on these default alignments, global advising alone, and local combined with global advising. Here the combined method uses local advising on all alternate alignments within global advising. The oracle set of cardinality $k = 10$ was used for both global and local advising.

Local advising greatly improves the alignment accuracy of default alignments (left two bars in each group). In the two most difficult benchmark bins (to the left of the figure) using local advising increases the average accuracy by 11.5% and 9.1% respectively. The accuracy increases on all bins. Overall using local advising increases the accuracy of the default alignments by an average of 4.5% across bins.

Combining local and global advising greatly improves the accuracy over either of the methods individually. This is most pronounced for the hardest to align benchmarks. For the bottom two bins, using both parameter advising and adaptive local realignment increases the accuracy by 23.0% and 25.6% over using just the default parameter choices. Additionally, using adaptive local realignment increases the accuracy by 5.9% on the bottommost bins over using parameter advising alone. On average thats an 8.9% increase in accuracy over all bins by using the combined

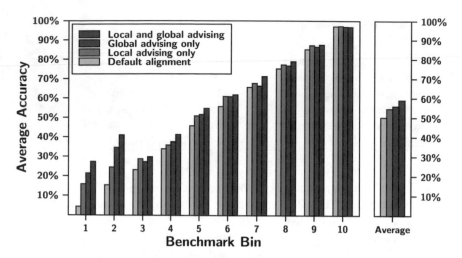

Fig. 8.3 Accuracy of the default alignment, local realignment of the default alignment, global advising, and global advising with local realignment, within difficulty bins. In the bar chart on the left, the horizontal axis shows all ten benchmarks bins, and the vertical bars show the accuracy averaged over just the benchmarks in that bin. The accuracy of the default alignment and parameter advising using an oracle set of cardinality $k = 10$, before local realignment is shown as well as the application of local realignment to both results. The bar chart on the right shows the accuracy uniformly averaged over the bins.

procedure over using just the default parameter choice, and a 3.1% increase over using only parameter advising.

8.3.2 Varying Advising Set Cardinality

The previous section focused on using advising sets of cardinality $k = 10$. Because an alignment is produced for each region of local realignment for each parameter choice in the advising set, it may be desirable to use a smaller set to reduce the running time of local (or or global) advising. Oracle advising sets were produced for cardinalities $k = 2, \ldots, 15$ and used to test the effect of local advising both alone and in combination with global advising. Figure 8.4 shows the average advising accuracies of using advising sets of increasing cardinalities under the three conditions described above as well as the combination of local with global advising where the local advising step is only performed on the single best alignment identified by global advising. The cardinality of the set used for both parameter advising and local realignment is shown on the horizontal axis, while the vertical axis shows the alignment accuracy of the produced alignments averaged first within difficulty bins then across bins.

The accuracy of alignments produced by all four methods eventually reaches a plateau where adding additional parameters to the advising set no longer increases

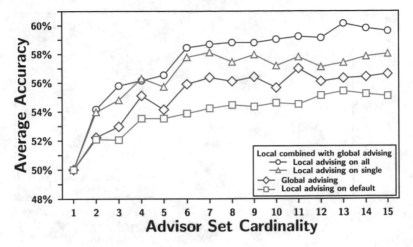

Fig. 8.4 Advising accuracy using various methods versus set cardinality. This figure compares the accuracy of alignments produced by local advising on the alignment produced using the Opal default parameter setting, global advising alone, and two variants combining local and global advising. The horizontal axis represents an increasing oracle advising set cardinality, used for both parameter advising and local realignment. The vertical axis shows the accuracy of the alignments produced by each of the advising methods averaged across difficulty bins.

the alignment accuracy. This plateau is reached at cardinality $k = 10$ when local realignment is applied to the default alignments, and at $k = 6$ for parameter advising with and without local realignment, but this plateau is higher for the combined methods.

Across all cardinalities, using local combined with global advising improves alignment accuracy by nearly 4% on average.

The results above give average advising accuracy uniformly weighted across *bins*. If the average accuracy is instead uniformly weighted across *benchmarks*, the default parameter choice of the Opal aligner achieves accuracy 80.4%. Applying both local and global advising at cardinality $k = 10$, this increases to 83.1% (performing local advising on all global alternate alignments). Using only local or global advising achieves accuracy 82.1% or 81.8% respectively. At $k = 5$ the accuracy of using local and global advising is 82.7%. By comparison, the average accuracy of other standard aligners on these benchmarks is: Clustal Omega, 77.3%; Muscle, 78.09%; MAFFT, 79.38%.

8.3.3   Comparing Estimators for Local Advising

The Facet estimator has been shown to give the best performance for the task of global advising compared to the other accuracy estimators available (see Chap. 6). Figure 8.5 shows the average accuracy of local advising on default alignments using

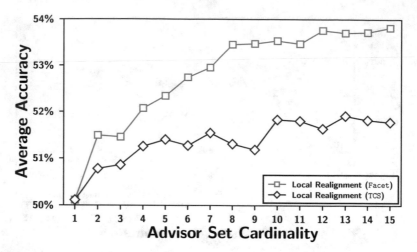

Fig. 8.5 Accuracy of the default alignment and local realignment using TCS and Facet with various advisor set cardinalities. This figure compares the accuracy of alignments produced by the Opal default parameter setting applying local realignment using either the TCS or Facet estimator. The horizontal axis represents an increasing oracle advising set cardinality used for local realignment. The vertical axis shows the accuracy of the alignments produced by each of the advising methods averaged across difficulty bins.

both Facet and TCS (the next best estimator for advising) using advisor sets of cardinality $k = 2, \ldots, 15$. Using TCS for local advising greatly increased the running time because it is an additional system call and additional file IO. Because of the additional computational requirements iteration was not performed on local advising for either estimator. Using TCS for local advising gives an increase in accuracy of less than half that of Facet.

8.3.4 Effect of Iterating Local Realignment

The local advising process can be considered a refinement step for multiple sequence alignment. To continue refining the alignment we can iterate the local advising procedure (see Sect. 8.2.3). Iterating local advising should eventually converge with no further improvement in the result (i.e. the alignment regions are not being changed) or even worse deteriorate due to noise in the accuracy estimator. To find the optimal iteration limit all iteration limits from 1 to 25 were tested. The peak accuracy on training benchmarks was at five iterations, and that value was used for all other experiments shown in this chapter (other than those in Sect. 8.3.3). The table below shows the average accuracy of using local adaptive realignment on the default alignment with various numbers of iterations.

Iterations	1	2	3	4	5	10	15	25
Testing	53.5%	53.7%	54.1%	54.4%	54.5%	54.5%	54.5%	54.5%
Training	53.5%	53.9%	54.5%	54.6%	54.8%	54.8%	54.9%	54.9%

Table 8.2 Summary of local realignment on default alignments

Bin	1	2	3	4	5	6	7	8	9	10	Overall
Total number of benchmarks	12	12	20	34	26	50	61	74	137	434	861
Benchmarks unchanged	4	5	4	7	7	16	16	13	22	82	176
Benchmarks modified by adaptive local realignment	8	7	16	27	19	34	46	61	115	352	685
Percentage of benchmarks altered	67%	58%	80%	79%	73%	68%	74%	82%	84%	81%	80%
Average *Bad* regions per benchmark	1.92	2.17	2.50	1.88	2.23	2.14	2.31	2.16	2.48	2.19	2.23
Average percentage of original column realigned	75%	73%	76%	70%	75%	77%	74%	73%	75%	72%	73%
Average percentage of original column replaced	64%	60%	68%	60%	66%	72%	65%	63%	64%	47%	57%

8.3.5 Summarizing the Effect of Adaptive Local Realignment

Table 8.2 summarizes how adaptive local realignment behaves across difficulty bins when used to modify alignments produced using the default parameter setting in Opal. The first two rows show how many of the 861 benchmarks are in each bin, as well as how many of them had at least one realignment region where the advisor chose to replace the global alignment. The fourth row shows the average number of *Bad* regions in a benchmark; on average about two regions were realigned for each default alignment. The last two rows summarize the percentage of the original columns those *Bad* regions covered, and how many of the columns from the original alignment ended up being replaced. In the easiest-to-align benchmark bin only 47% of the alignment columns were altered, while in the rest of the bins over 60% of the alignment columns were improved.

8.3.6   Running Time

As currently implemented in Opal, local advising does not take advantage of the
independence of the calls to the aligner in the parameter advising step and run
them in parallel. Therefore a large increase in time consumption is seen when
generating locally advised alignments. In particular the average time for computing
an alignment using the default global parameter setting goes from about 8 sec to
just over 36 sec using an advisor set cardinality of $k = 10$. When iterating the local
advising step five times we see the average running time increase to 110 sec.

In contrast *global* advising exploits the independence of the aligner on different
parameter settings. The running time for advisor set cardinality $k = 10$ for global
advising alone is around 33 sec, much less than the tenfold increase to be expected
if advising was not done in parallel. Even though global advising is done in parallel,
local advising is not; the average running time over all benchmarks increases to 68
and 178 sec for combining local and global advising by performing local advising
on all global alignments with and without iteration, respectively.

8.3.7   Local and Global Advising in Opal

The development trunk of the Opal aligner includes the ability to perform adaptive
local realignment both with and without parameter advising. To achieve the same
results shown here the following commands were used to run the aligner:

(A) **Parameter advising with local realignment on all**
```
java opal.Opal --in <input file>\
   --facet_structure <structure file>\
   --configuration_file <parameter set>\
   --out_best <output file>
```

(B) **Parameter advising with local realignment on single**
```
java opal.Opal --in <input file>\
   --facet_structure <structure file>\
   --configuration_file <parameter set>\
   --out_prealignment_best_realignment\
      <output file>
```

(C) **Parameter advising without local realignment**
```
java opal.Opal --in <input file>\
   --facet_structure <structure file>\
   --advising_configuration_file <set>\
   --out_best <output file>
```

(D) **Default alignment with local realignment**
```
java opal.Opal --in <input file>\
  --facet_structure <structure file>\
  --realignment_configuration_file <set>\
  --out <output file>
```

(E) **Default alignment without local realignment**
```
java opal.Opal --in <input file>\
  --out <output file>
```

Summary

This chapter has described *adaptive local realignment*, the first method that demonstrably boosts protein multiple sequence alignment accuracy by locally realigning regions that may have distinct mutation rates using different aligner parameter settings. Applying this new method alone to alignments initially computed using a single optimal default parameter setting already improves alignment accuracy significantly. When combined with methods to select an initial non-default parameter setting for the particular input sequences through global parameter advising, this new local parameter advising method greatly improves accuracy even further. Software that performs adaptive local realignment is freely available within the Opal aligner.

Chapter 9
Core Column Prediction for Alignments

In a computed multiple sequence alignment, the *coreness* of a column is the fraction of its substitutions that are in so-called core columns of the unknown reference alignment of the sequences, where the core columns of the reference alignment are those that are reliably correct. In the absence of knowing the reference alignment, the coreness of a column can only be estimated. This chapter describes the first method for estimating column coreness for protein multiple sequence alignments.

The approach to predicting coreness is similar to nearest-neighbor classification from machine learning, except nearest-neighbor distances are transformed into a coreness estimate using a regression function, and an appropriate distance function is automatically learned through a new optimization formulation that solves a large-scale linear programming problem. The included experiments apply the coreness estimator to improving *parameter advising*, the task of choosing good parameter values for an aligner's scoring function, and show that our estimator strongly outperforms others from the literature, providing a significant boost in advising accuracy.

9.1 Column Coreness

The accuracy of a multiple sequence alignment computed on a benchmark set of input sequences is usually measured with respect to a *reference alignment* that represents the gold-standard alignment of the sequences. For protein sequences, reference alignments are typically determined by structural superposition of the known three-dimensional structures of the proteins in the benchmark. The accuracy of a computed alignment is then defined to be the fraction of substitutions of pairs of residues in the so-called *core columns* of the reference alignment that are also

---

Adapted from publications [27, 29].

© Springer International Publishing AG 2017

D. DeBlasio, J. Kececioglu, *Parameter Advising for Multiple Sequence Alignment*,
Computational Biology 26, https://doi.org/10.1007/978-3-319-64918-4_9

present in columns of the computed alignment. Core columns represent those in the reference that are deemed to be reliable, and are columns containing a residue from every input sequence such that the pairwise distances between these residues in the structural superposition of the proteins are all within some threshold (typically a few angstroms). In short, given a known reference alignment whose columns are labeled as either core or non-core, the accuracy of any other computed alignment of its proteins can be determined by evaluating the fraction of substitutions in these core columns that are recovered. For a given column in a computed alignment, the *coreness* value of the column is determined to be the fraction of its substitutions that are in core columns of the reference alignment. A coreness value of 1 means the column of the computed alignment corresponds to a core column of the reference alignment.

When aligning sequences in practice, obviously such a reference alignment is not known, and the accuracy of the computed alignment, or the coreness of its columns, must be estimated. As the preceding chapters of this book have already shown, a good *accuracy estimator* for computed alignments is extremely useful. It can be leveraged to: pick among alternate alignments of the same sequences the one of highest estimated accuracy, for example, to choose good parameter values for an aligner's scoring function, called *parameter advising*; or to select the best result from a collection of different aligners, yielding a natural *ensemble aligner* that can be far more accurate than any individual aligner in the collection.

Similarly, a good *coreness estimator* for columns in a computed alignment can be used to: mask out unreliable regions of the alignment before computing an evolutionary tree; or to improve an alignment accuracy estimator by concentrating its evaluation function on columns of higher estimated coreness, thereby boosting the performance of parameter advising. In fact, in principle a perfect coreness estimator would itself yield an ideal accuracy estimator.

In this chapter, we describe the first column-coreness estimator for protein multiple sequence alignments. This approach to predicting coreness is similar in some respects to nearest-neighbor classification from machine learning, except the nearest-neighbor distance is transformed into a coreness estimate using a regression function, and we automatically learn an appropriate distance function through a new optimization formulation that solves a large-scale linear programming problem. The performance of this new coreness estimator is evaluated by applying it to the task parameter advising in multiple sequence alignment.

Related Work

The method described in this chapter is the first fully general attempt to directly estimate the coreness of columns in computed protein alignments. In the literature, the GUIDANCE tool [92] gives reliability values for alignment columns, which they evaluate by measuring the classification accuracy of predicting totally correctly aligned core columns from reference alignments, though they do not attempt to relate reliability to coreness. GUIDANCE also requires alignments to contain at least four sequences, which limits the alignment benchmarks that can be considered. Tools are also available that assess the quality of columns in a multiple alignment, and can be categorized into those that compute a column quality score which can be thresholded, and those that only identify columns that are unreliable (for removal from further analysis). The popular quality score tools are TCS [18], ZORRO [112], and Noisy [35]; these can be used to modify the feature functions in an accuracy estimator such as Facet [25], as is later shown in Sect. 9.3.2. Tools that simply mask unreliable columns of an alignment include ALISCORE [62], GBLOCKS [17], and TrimAL [15].

The studies in this chapter focus on comparing our coreness estimator to TCS and ZORRO, as these are the most recent tools that provide quality scores, as opposed to simply masking columns. Furthermore, of the above tools, ALISCORE, GBLOCKS and GUIDANCE have been shown to be dominated by ZORRO, while Noisy in turn has been shown to be dominated by GUIDANCE.

9.2 Learning a Coreness Estimator

To describe how to learn a column coreness estimator, we first discuss the *representation* of alignment columns, and our grouping of consecutive columns into *window classes*; we then present the *regression function* for estimating coreness, which transforms a distance to a window class into a coreness value; and finally, we describe how to learn this window distance function by solving a large-scale *linear programming* problem.

9.2.1 Representing Alignment Columns

The representation of a multiple alignment column should be in a form that captures the association of amino acids and predicted secondary-structure types, but is independent of the number of sequences in the alignment. This is necessary for the labeled column examples in our training set to be useful for estimating the coreness of columns that come from other alignments with arbitrary numbers of sequences.

Let Σ be the 20-letter amino acid alphabet, and $\Gamma = \{\alpha, \beta, \gamma\}$ be the secondary-structure alphabet, corresponding respectively to types α-helix, β-strand, and *other* (also called *coil*). The association of an amino acid $c \in \Sigma$ is encoded along with its predicted secondary structure type $s \in \Gamma$ using an ordered pair (c, s) that is called a *state*, from the set $Q = (\Sigma \times \Gamma) \cup \{\xi\}$. Here $\xi = (\varepsilon, \varepsilon)$ is the *gap state*, where $\varepsilon \notin \Sigma$ is the alignment *gap character* (often displayed by the dash symbol '-').

A multiple alignment column can then be represented as a distribution over the set of states Q, which is called its *profile* (mirroring standard terminology

[36, p. 101]). The profile C for a given column is denoted by a function $C(q)$ on states $q \in Q$ satisfying $C(q) \geq 0$ and $\sum_{q \in Q} C(q) = 1$. For a column $(c_1 c_2 \cdots c_k)$ in a multiple alignment of k sequences, with associated predicted secondary structure types $(s_1 \cdots s_k)$, where for a gap $c_i = \varepsilon$ the associated secondary structure type is also $s_i = \varepsilon$, its profile C is,

$$ C(q) \; := \; \frac{1}{k} \left| \{i \, : \, (c_i, s_i) = q\} \right| . $$

In other words, $C(q)$ is the relative frequency of state q in the column.

This is generalized to secondary structure predictions that for amino acid c_i give *confidences* $p_i(\alpha), p_i(\beta), p_i(\gamma)$ that the amino acid is in each of the three secondary structure types (where these confidences sum to 1), as follows. For state $q = (a, s) \neq \xi$, profile C is then,

$$ C(q) \; := \; \frac{1}{k} \sum_{1 \leq i \leq k \, : \, c_i = a} p_i(s) . $$

In other words, $C(q)$ is now the normalized total confidence in state $q \neq \xi$. For gap state $q = \xi$, value $C(\xi)$ is the same as before.

9.2.2   Classes of Column Windows

The ground truth of whether a column in a reference alignment is core or non-core depends on whether the residues of the proteins in that column are sufficiently close in space in the structural superposition of the folded 3-dimensional structures of the proteins. This folded structure at a residue is not simply a function of the amino acid of the residue itself, or its secondary structure type, but is also a function of the nearby residues in the protein. Consequently, to estimate the coreness of a given column in a computed alignment, additional contextual information is needed which can be extracted from nearby columns of the alignment.

This additional context around a given column is collected by forming a window of consecutive columns centered on the given column. Formally, a *column window* W of width $w \geq 1$ is a sequence of $2w + 1$ consecutive column profiles $W_{-w} \cdots W_{-1} W_0 W_{+1} \cdots W_{+w}$ centered around profile W_0.

A set of *window classes* $\mathcal{C}$ is defined based on the columns in a labeled training window which are known to be core or non-core with respect to their reference alignment. A column labeled core is denoted by C, and a column labeled non-core by N. For window width $w = 1$, such labeled windows can be described by strings of length 3 over alphabet $\{C, N\}$. The three classes of *core windows* are CCC, CCN, NCC, and the three classes of *non-core windows* are CNN, NNC, NNN. (A window is considered core or non-core depending on the label of its center column.) Together these six classes comprise set $\mathcal{C}$. The five classes with at least one core column C in the window are called *structured classes*, and the one class with no core columns is the single *unstructured class*, which is denoted by the symbol $\perp = \text{NNN}$.

Reference alignments explicitly label their columns as core or non-core. For computed alignments, which have a known reference alignment, a column is labeled as core or non-core depending on whether the true coreness value for the column is above a fixed threshold.

9.2.3 The Coreness Regression Function

An estimator for the coreness of a column is learned by fitting a regression function that first measures the similarity between a window around the column and training examples of windows with known coreness, and then transforms this similarity into a coreness value.

The similarity between windows is expressed in terms of the similarity of their corresponding columns. A *distance function d* on pairs of column profiles A, B is used to measure the similarity between columns and has the form

$$d(A, B) := \sum_{p,q \in Q} A(p)\, B(q)\, \sigma(p, q),$$

where $\sigma(p, q)$ is a substitution score that measures the dissimilarity between the pair of states p, q.

This is then extended to a distance d on pairs of windows $V = V_{-w} \cdots V_w$ and $W = W_{-w} \cdots W_w$ by,

$$d(V, W) := \sum_{-w \le i \le +w} d_i(V_i, W_i),$$

where the d_i are *positional* distance functions on column profiles. Function d_i is given by its *positional* substitution scores $\sigma_i(p, q)$. The positional σ_i can score dissimilarity higher at positions i at the center of the window, and lower toward the edge of the window.

Finally, this is extended to *class-specific* window distance functions d_c that are specific to each window class $c \in C - \{\perp\}$. Function d_c is given by its class-specific positional profile distance functions $d_{c,i}$, which are in turn given by class-specific positional substitution scores $\sigma_{c,i}$.

The *regression function* that estimates the coreness of a column first forms a window W centered on the column, and then performs the following. To transform a distance to coreness we use two different functions: function f_{core} for core classes, and function f_{non} for non-core classes.

(1) (*Find distance to closest class*) Across all labeled training windows, in all structured window classes, find the training window that has smallest class-specific distance to W. Call this closest window V, its class c, and their distance $\delta = d_c(V, W)$.

(2) (*Transform distance to coreness*) If class c is a core class, return coreness value $f_{core}(\delta)$. Otherwise, return value $f_{non}(\delta)$.

9.2.3.1 Finding the Distance to a Class

Finding the distance of a window W to a class c, requires first finding the *nearest neighbor* of W among the set of training windows T_c in class c, namely $\operatorname{argmin}_{V \in T_c}\{d_c(V, W)\}$. Finding the nearest neighbor through exhaustive search by explicitly evaluating $d_c(V, W)$ for every window V can be expensive when T_c is large (and cannot be avoided in the absence of exploitable properties of function d_c).

When the distance function is a *metric*, for which the key property is the *triangle inequality* (namely that $d(x, z) \le d(x, y) + d(y, z)$ for any three objects x, y, z), faster nearest neighbor search is possible. In this situation, a preprocessing step first builds a data structure over the set T_c, which then allows for faster nearest neighbor searches on T_c for any query window W. One of the best data structures for nearest neighbor search under a metric is the *cover tree* of Beygelzimer, Kakade and Langford [11]. Theoretically, cover trees permit nearest neighbor searches over a set of n objects in $O(\log n)$ time, after constructing a cover tree in $O(n \log n)$ time, assuming that the intrinsic dimension of the set under metric d has a so-called bounded expansion constant [11]. (For actual data, the expansion constant can be exponential in the intrinsic dimension.) In the experiments later in this chapter, the *dispersion tree* data structure of Woerner and Kececioglu [111] is used, which in extensive testing on scientific data is significantly faster in practice than cover trees.

A separate dispersion tree is produced for each structured window class $c \in C - \{\bot\}$ over its training set T_c using its distance function d_c, in a preprocessing step. To find the nearest neighbor to window W over all training windows $T = \bigcup_c T_c$, a nearest neighbor search is performed with W in each class dispersion tree, and we merge these $|C| - 1$ search results by picking the one with smallest distance to W.

9.2.3.2 Transforming Distance to Coreness

A sigmoid function is used to transform nearest neighbor distance into a coreness value. Once the distance functions d_c are learned, as described in Sect. 9.2.4, the transform function is fit to empirical coreness values measured at the distances observed for example windows from our set of training windows. The examples are sorted by their observed nearest neighbor distance, and at each observed distance δ, the m adjacent examples whose distances are below δ are then collected, along with the m adjacent examples above δ. The average true coreness value of these $2m + 1$ examples is then computed, and this average true coreness value is assigned to distance δ. A sigmoid curve is then fit to these pairs of average true coreness and observed nearest neighbor distance values. This fitting process is performed separately for example windows from core classes, and non-core classes.

The particular sigmoid that we fit is the *logistic function*. The general form of the logistic function f that we use is,

$$f(x) \;:=\; \ell \;+\; (u - \ell)\, \frac{1}{1 + e^{ax+b}},$$

with four parameters a, b, ℓ, u, where ℓ and u are respectively the minimum and maximum average true coreness values observed for the examples, and a and b are shape parameters. The curve fitting tools in `SciPy` [52] (which are a wrapper for `MINPACK` [73]) were used to find values for the shape parameters a, b that best fit the data.

Logistic functions $f_{\mathrm{core}}(\delta)$ and $f_{\mathrm{non}}(\delta)$ are fit separately, with their own parameter values a, b, ℓ, u, to data from the core and non-core classes, respectively. For function f_{core}, shape parameter a is positive (so coreness is decreasing in the distance δ to a *core* class); for f_{non}, parameter a is negative (so coreness is increasing in the distance δ from a *non-core* class). As Fig. 9.1 in Sect. 9.4.1.2 later shows, these logistic transform functions fit actual coreness data remarkably well.

9.2.4 Learning the Distance Function by Linear Programming

This section describes the linear program used to learn the distance function on column windows. The linear program learns a different, *class-specific*, distance

Fig. 9.1 Fit of the logistic transform functions for the coreness regressor to the average true coreness of training examples at each nearest neighbor distance.

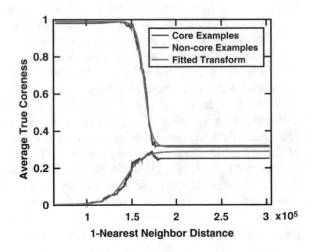

function d_c for each window class $c \in C$. These distance functions d_c are made commensurate between classes by a final rescaling step after solving the linear program.

Again the window classes C are divided into two categories: the *structured classes*, containing windows centered on core columns, or centered on non-core columns that are flanked on at least one side by core columns; and the *unstructured class*, containing windows of only non-core columns. The unstructured class of completely non-core windows is again denoted by $\perp \in C$.

In principle, the linear program tries to find distance functions d_c that would make the following "conceptual" nearest neighbor classifier accurate. (Note such a classifier is not what is being learned.) This conceptual classifier forms a window W centered on the column to be classified, and first finds the nearest neighbor to W over all structured classes $C - \{\perp\}$ in the training set using their corresponding distance functions d_c. If the distance to this nearest neighbor is at most a threshold τ, the central column of window W is declared "core" or "non-core" depending on whether this nearest structured class c is core or non-core. Otherwise, the nearest neighbor distance exceeds threshold τ, the window is deemed to be in the unstructured non-core class $\perp$, and its central column is declared "non-core." The key aspect of this conceptual nearest neighbor classifier is that it can recognize a *completely non-core* window W from class $\perp$, without actually having any examples in its training set that are close to W. This is critical for our coreness estimation task, as the set of possible windows from the unstructured class $\perp$ is enormous and probably lacks any recognizable structure, which makes identifying them through having a near neighbor in the training set essentially hopeless. On the other hand, identifying windows from the structured classes is possible by having sufficiently many examples in the training set. The following linear program learns both distance functions d_c and such distance thresholds τ_c.

To construct the linear program, the *training set* $\mathcal{T}$ of labeled windows by window class is partitioned as follows: subset $T_c \subseteq \mathcal{T}$ contains all training windows of class $c \in C$. A smaller *training sample* $S_c \subseteq T_c$ is formed for each class c by choosing a random subset of T_c with a specified cardinality $|S_c|$.

The *constraints* of the linear program fall in several categories. For a sample training window $W \in S_c$, other windows $V \in T_c$ from the same class c in the full training set that are close to W are identified (under a default distance $\widetilde{d}_c$). These close windows V from the same class c are called *targets*. Similarly for $W \in S_c$, other windows $U \in T_b$ from a different class $b \neq c$ in the full training set that are also close to W are also identified (under $\widetilde{d}_b$). These other close windows U from a different class b are called *impostors*. (This parallels the terminology of Weinberger and Saul [105].) More formally, the *neighborhood* $\mathcal{N}_c(W, i)$ for a structured class $c \in C - \{\perp\}$ denotes the set of i-nearest-neighbors to W (not including W) from training set T_c under the class-specific *default distance* function $\widetilde{d}_c$. (The default distance function that we use in our experiments is described in Sect. 9.4.1.1.) The constraints of the linear program find distance functions that for a sample window $W \in S_c$, *pull in* targets $V \in \mathcal{N}_c(W, i)$ by making $d_c(V, W)$ small, and *push away* impostors $U \in \mathcal{N}_b(W, i)$ for $b \neq c$ by making $d_b(U, W)$ large.

The neighborhoods $\mathcal{N}(W, i)$ that give the sets of targets and impostors for the linear programming formulation are defined with respect to default distance functions $\tilde{d}$, as mentioned above. These neighborhoods really should be defined with respect to the learned distance functions d_c, but obviously they are not available until after the linear program is solved. This discrepancy is addressed by iteratively solving a series of linear programs. The first linear program at iteration 1 defines neighborhoods with respect to distance functions $d^{(0)} = \tilde{d}$, and its solution yields the new functions $d^{(1)}$. In general, iteration i uses the previous iteration's functions $d^{(i-1)}$ to formulate a linear program whose solution yields the new distance functions $d^{(i)}$. This process is repeated for a fixed number of iterations, or until the change in the distance functions is sufficiently small.

The *target constraints* for each sample window $W \in S_c$ from each structured class $c \in \mathcal{C} - \{\bot\}$, and each target window $V \in \mathcal{N}_c(W, k)$, are,

$$e_{VW} \geq d_c(V, W) - \tau_c, \tag{9.1}$$

$$e_{VW} \geq 0, \tag{9.2}$$

where e_{VW} is a target *error variable* and τ_c is a *threshold variable*. In the above, quantity $d_c(V, W)$ is a linear expression in the *substitution score variables* $\sigma_{c,i}(p, q)$, so constraint (9.1) is a linear inequality in all these variables. Intuitively, if condition $d_c(V, W) \leq \tau_c$ holds (so W will be considered to be in its correct class c), in the solution to the linear program, variable e_{VW} will equal $\max\{d_c(V, W) - \tau_c, 0\}$, the amount of error by which this ideal condition is violated.

In the target neighborhood $\mathcal{N}_c(W, k)$ above, parameter k specifies the number of targets for each sample window W. In the experiments later a small number of targets are used, with $k = 2$ or 3.

The *impostor constraints* for each sample window $W \in S_c$ from each structured class $c \in \mathcal{C} - \{\bot\}$, and each impostor window $V \in \mathcal{N}_b(W, \ell)$ from each structured class $b \in \mathcal{C} - \{\bot\}$ with $b \neq c$, are,

$$f_W \geq \tau_b - d_b(V, W) + 1, \tag{9.3}$$

$$f_W \geq 0, \tag{9.4}$$

where f_W is an impostor error variable. Intuitively, if condition $d_b(V, W) > \tau_b$ holds (so W will not be considered to be in the incorrect class b), this can be expressed by $d_b(V, W) \geq \tau_b + 1$ using a *margin* of 1. (Since the scale of the distance functions is arbitrary, a unit margin can always be picked without loss of generality.) In the solution to the linear program, variable f_W will equal $\max_{b \in \mathcal{C} - \{\bot\}, V \in \mathcal{N}_b(W, \ell)} \{\tau_b - d_b(V, W) + 1, 0\}$, the largest amount of error by which this condition is violated for W across all b and V.

Impostor constraints are also created for each completely non-core window $W \in T_\bot$, and each core window $V \in \mathcal{N}_b(W, \ell)$ from each structured core class b (as W should not be considered core), which are of the same form as inequalities (9.3) and (9.4) above.

In the impostor neighborhood $\mathcal{N}_b(W, \ell)$ above, parameter ℓ specifies the number of impostors for each sample window W. A large number of impostors $\ell \approx 100$ are used in the later experiments. Having a single impostor error variable f_W per sample window W (versus a target error variable e_{VW} for every W and target V) allows us to use a very large ℓ while still keeping the number of variables in the linear program tractable.

The *triangle inequality constraints*, for each structured class $c \in C - \{\perp\}$, each window position $-w \leq i \leq w$, and all states $p, q, r \in Q$ (including the gap state ξ), are,

$$\sigma_{c,i}(p, r) \;\leq\; \sigma_{c,i}(p, q) \;+\; \sigma_{c,i}(q, r). \tag{9.5}$$

These reduce to simpler inequalities when states p, q, r are not all distinct or coincide with the gap state (which are not enumerated here).

The remaining constraints, for all classes $c \in C$, positions $-w \leq i \leq w$, states $p, q \in Q$, and gap state ξ, are,

$$\sigma_{c,i}(p, q) \;=\; \sigma_{c,i}(q, p), \tag{9.6}$$

$$\sigma_{c,i}(p, p) \;\leq\; \sigma_{c,i}(p, q), \tag{9.7}$$

$$\sigma_{c,i}(p, q) \;\geq\; 0, \tag{9.8}$$

$$\sigma_{c,i}(\xi, \xi) \;=\; 0, \tag{9.9}$$

$$\tau_c \;\geq\; 0, \tag{9.10}$$

which ensure the distance functions are symmetric and non-negative. (We do not enforce the other metric conditions $d_c(W, W) = 0$ and $d_c(V, W) > 0$ for $V \neq W$, as these are not needed for our coreness estimation task, and we prefer having a less constrained distance d_c that might better minimize the following error objective.)

Finally, the *objective function* minimizes the average error over all training sample windows. Formally, we minimize,

$$\alpha \; \frac{1}{|C| - 1} \sum_{c \in C - \{\perp\}} \frac{1}{|S_c|} \sum_{W \in S_c} \frac{1}{k} \sum_{V \in \mathcal{N}_c(W, k)} e_{VW}$$

$$+ \; (1 - \alpha) \frac{1}{|C|} \sum_{c \in C} \frac{1}{|S_c|} \sum_{W \in S_c} f_W,$$

where $0 \leq \alpha \leq 1$ is a blend parameter controlling the weight on target error versus impostor error. Note that in an optimal solution to this linear program, $e_{VW} = \max\{d_c(V, W) - \tau_c, 0\}$ and $f_W = \max_{V,b}\{\tau_b - d_b(V, W) + 1, 0\}$, since inequalities (9.1)–(9.4) ensure the error variables are at least these values, while minimizing the above objective function ensures they will not exceed them. Thus solving the linear program finds distance functions d_c, given by substitution scores $\sigma_{c,i}(p,q)$, that minimize the average over the training windows $W \in S_c$ of the amount of violation of our ideal conditions $d_c(V, W) \leq \tau_c$ for targets $V \in T_c$ and $d_b(V, W) > \tau_b$ for impostors $V \in T_b$.

To summarize, the variables of the linear program are the substitution scores $\sigma_{c,i}(p,q)$, the error variables e_{VW} and f_W, and the threshold variables τ_c. For n total training sample windows, k impostors per sample window, m window classes of width w, and amino acid alphabet size s, this is $\Theta(kn + s^2 wm)$ total variables. The main constraints are the target constraints, impostor constraints, and triangle inequality constraints. For ℓ impostors per sample window, this is $\Theta((k + \ell m)n + s^3 wm)$ total constraints. To ensure that solving the linear program is tractable the number ℓ of impostors is carefully defined as is the total size n of the training sample.

After solving the linear program, we *rescale* the distance functions so their corresponding distance thresholds all match the common value $\tau := \max_{c \in C} \tau_c$. Specifically, each distance function d_c is scaled up by multiplying its substitution scores $\sigma_{c,i}(p,q)$ (and distance threshold) by factor τ/τ_c. (Scaling up maintains that each class has a margin of at least 1.) This makes the distance functions d_c commensurate across classes. A conceptual 1-nearest-neighbor classifier for window W (which is not employed) could then just find the nearest neighbor of W across all structured classes using their class-specific distance functions, say it is window V from class c, and classify W as a member of structured class c if $d_c(V, W) \leq \tau$, and as a member of the unstructured non-core class $\perp$ otherwise. In actuality, rather than classifying W, its 1-nearest-neighbor distance $d_c(V, W)$ is mapped to a coreness value, as described in Sect. 9.2.3.2.

9.2.4.1 Ensuring the Triangle Inequality

This section shows that the resulting distance functions satisfy the triangle inequality, which allows for the use of fast data structures for metric-space nearest-neighbor search when evaluating the coreness estimator (as discussed in Sect. 9.2.3.1).

Theorem 9.1 (Triangle Inequality on Window Distance) *The class distance functions d_c obtained by solving the linear program satisfy the triangle inequality.*

Proof. For every class c, and all windows U, V, and W,

$$d_c(U,W) = \sum_i \sum_{p,r} U_i(p)\, W_i(r)\, \sigma_{c,i}(p,r)$$

$$= \sum_i \sum_{p,q,r} U_i(p)\, V_i(q)\, W_i(r)\, \sigma_{c,i}(p,r) \tag{9.11}$$

$$\leq \sum_i \sum_{p,q,r} U_i(p)\, V_i(q)\, W_i(r) \left(\sigma_{c,i}(p,q) + \sigma_{c,i}(q,r) \right) \tag{9.12}$$

$$= \sum_i \sum_{p,q} U_i(p) \ V_i(q) \ \sigma_{c,i}(p,q)$$

$$+ \sum_i \sum_{q,r} V_i(q) \ W_i(r) \ \sigma_{c,i}(q,r) \qquad\qquad (9.13)$$

$$= \sum_i d_{c,i}(U_i, V_i) + \sum_i d_{c,i}(V_i, W_i)$$

$$= d_c(U,V) + d_c(V,W),$$

where Eq. (9.11) follows from $\sum_q V_i(q) = 1$; inequality (9.12) follows from constraint (9.5) in the linear program; and Eq. (9.13) follows from $\sum_r W_i(r) = \sum_p U_i(p) = 1$. $\qquad\qquad\qquad\qquad\qquad\qquad\qquad\qquad\qquad\qquad\qquad\qquad\square$

9.3   Using Coreness to Improve Accuracy Estimation

Recall from Chap. 3 that the Facet accuracy estimator is a linear combination of efficiently-computable feature functions that are positively correlated with the true accuracy of an alignment. In general, the true accuracy of a computed alignment is evaluated just with respect to the columns of the reference alignment that are labeled as core; non-core columns do not contribute to true accuracy. Consequently, the ability to predict whether a column in a computed alignment corresponds to a core column in the unknown reference, or even better, to predict the coreness value of the column, should afford improved feature functions. The predicted coreness of computed alignment columns is used to improve the Facet estimator by: (1) creating a new feature function that attempts to directly estimate alignment accuracy by essentially counting the number of columns in the computed alignment that are predicted to be core and dividing by the estimated number of core columns in the reference, and (2) modifying the original feature functions so their evaluation is concentrated on columns with high predicted coreness.

9.3.1   Creating a New Coreness Feature

The alignment accuracy measure known in the literature as "total column score" (or TC-score) is defined as the number of core columns in the reference alignment that are perfectly aligned in the computed alignment, divided by the number of core columns in the reference. The feature function called *Predicted Alignment Coreness* is designed to estimate the total column score of a computed alignment (which cannot be exactly determined as the correct reference alignment is unknown). Denote our *coreness estimator* from Sect. 9.2.3 for alignment column C by $\chi(C)$, which predicts coreness by evaluating our coreness regression function on a window centered on C. For a computed multiple sequence alignment $\mathcal{A}$ of a set of sequences $\mathcal{S}$, and a given coreness threshold κ, the Predicted Alignment Coreness feature function is,

$$F_{\text{AC}}(\mathcal{A}) \;\; := \;\; \frac{\left| \{ C \in \mathcal{A} \; : \; \chi(C) \geq \kappa \} \right|}{L(\mathcal{S})}, \tag{9.14}$$

where the numerator counts the number of columns of $\mathcal{A}$ whose predicted coreness is above threshold κ (in which case the column is effectively predicted as being core), and the normalization function L in the denominator is an estimate of the number of core columns in the unknown reference alignment of the sequences $\mathcal{S}$. The estimator L is a polynomial in several easily-computed quantities of sequences $\mathcal{S}$, whose coefficients are found by fitting L on benchmark sets of sequences for which a reference alignment (and the true number of core columns) is known.

9.3.1.1 Estimating the Number of Core Columns

Function $L(\mathcal{S})$ that estimates the number of core columns in the reference alignment should tend to be increasing in the length of the sequences, and decreasing in their dissimilarity. The form of the estimator that will be considered is a polynomial whose terms are generally the product of a measure of sequence length and a fractional quantity related to the percent identity of the sequences. A variety of such measures are considered, which gives a polynomial with many terms, and then solve a linear programming problem to find their coefficients by minimizing the L_1-norm to true core column counts on example benchmarks, which effectively selects the appropriate terms (since many coefficients turn out to be zero).

The length measures on sequence set $\mathcal{S}$ that are considered include the maximum, minimum, and average length of the sequences in set $\mathcal{S}$. The set $\mathcal{L}$ consist of these three length measures ℓ_{max}, ℓ_{min}, and ℓ_{avg}. The similarity measures on $\mathcal{S}$ that are considered include forms of percent identity, evaluated by summing over all pairs of sequences in $\mathcal{S}$ the maximum number of identities between each pair of sequences (computed by dynamic programming using the identity substitution matrix with no gap penalties), and normalizing by summing over all pairs of sequences the minimum, maximum, or average lengths of the sequences, giving percent identity measures p_{min}, p_{max}, and p_{avg}. The set $\mathcal{P}$ consists of these three percent identity measures. As a gap dissimilarity measure we also consider the difference in length between the longest and shortest sequences in $\mathcal{S}$ normalized by any of the length measures, giving the ratio measures r_{max}, r_{min}, and r_{avg}, as well as the length ratios $r_{\text{mm}} := \ell_{\text{min}}/\ell_{\text{max}}$, $r_{\text{am}} := \ell_{\text{avg}}/\ell_{\text{max}}$, $r_{\text{ma}} := \ell_{\text{min}}/\ell_{\text{avg}}$. Call $\mathcal{R}$ the set of these ratio measures.

The general form of estimator L is then,

$$L(S) := \sum_{\ell \in \mathcal{L}, p \in \mathcal{P}} c_{\ell p} \; \ell(S) \; p(S) \; + \sum_{\ell \in \mathcal{L}, r \in \mathcal{R}} c_{\ell r} \; \ell(S) \; r(S) \; +$$

$$\sum_{\ell \in \mathcal{L}, p \in \mathcal{P}, r \in \mathcal{R}} c_{\ell pr} \; \ell(S) \; p(S) \; r(S) \, .$$

Coefficients $c_{\ell p}, c_{\ell r}, c_{\ell pr}$ are fit by solving a linear program that minimizes the sum of the absolute values of the differences between the true number of core columns and the estimated number over all reference alignments in our suite of benchmarks.

The fitted function L that is used for evaluating the Predicted Alignment Coreness feature F_{AC} is given in Sect. 9.4.1.3.

9.3.2 Augmenting Former Features by Coreness

Since the true accuracy of a computed alignment is measured just with respect to the core columns of a reference alignment, and non-core columns are ignored, concentrating an accuracy estimator on columns with higher coreness should improve the estimator. Accordingly, the alignment feature functions used by the Facet estimator (see Chap. 3) are modified to focus their evaluation on columns of higher predicted coreness. This section discuss only those features that can incorporate coreness; a full description of all feature functions in Facet is in Chap. 3.

Secondary Structure Blockiness takes secondary structure predictions from PSIPRED [51] and finds a packing of secondary structure blocks of maximum total score, where a block is an interval of columns and a subset of the sequences such that all residues in the block have the same secondary structure prediction, a packing is a set of blocks whose column intervals are disjoint, and the score of a block is the total number of pairs of residues within the columns in the block. The score of a block is modified by weighting the number of pairs per column by the column's predicted coreness. *Secondary Structure Identity* is the fraction of substitutions in the computed alignment that share the same predicted secondary structure, which is modified by weighting the count of substitutions with shared structure by their column's predicted coreness. *Amino Acid Identity* uses predicted coreness to weight the fraction of substitutions in a column that are in the same amino acid equivalence class. *Average Substitution Score* is modified by averaging the BLOSUM62 score [48] of all substitutions, weighted by their column's predicted coreness.

9.4 Assessing the Coreness Prediction

In this section the same benchmark sets and weights are used that were described in Chap. 6. The same universe of parameter choices is also to conduct 12-fold cross validation experiments using the `Opal` aligner.

9.4.1 Constructing the Coreness Regressor

Results on constructing the coreness regressor are presented first. Specifically, result on learning its distance function, mapping distances to coreness, and estimating the number of core columns.

9.4.1.1 Learning the Distance Function

The set of column windows for each class were constructed using the reference alignments of the benchmarks in the training set for each cross-validation fold. A subsampling of 4000 examples of each class was put into the set of training examples, and 4000 examples (or the remaining examples of that class, whichever is smaller) are put into the database for each class searched for nearest neighbors. A subset of 2000 of the 24,000 collected training examples for learning distances was used to reduce the training time. A similar set of 2000 windows was collected from the alignments of testing benchmarks, to test the generalization of the distance functions when used for core column prediction.

A default distance between the training windows and each example window in the database for each class was used to get the initial sets of targets and impostors. The default distance on a pair of states is a linear combination of the VTML200 amino acid substitution score (shifted and scaled to a dissimilarity value in the range $[0, 1]$) and the identity of the secondary structure prediction. For each column i with $-1 \leq i \leq 1$ in a window of three columns, the column weight w_i were set to $w_0 = \frac{1}{2}$, $w_{\{-1,+1\}} = \frac{1}{4}$ for all columns in a class c that are core, and $w_i = 0$ for non-core columns. The distance between states $p = (a, s)$ and $q = (b, t)$ in the ith column of class c is,

$$\sigma_{c,i}(p, q) = w_i \left(\alpha \, \text{VTML200}(a, b) + (1 - \alpha) [s \neq t] \right),$$

where $\alpha = \frac{1}{2}$, and expression $[s \neq t]$ evaluates to 1 if $s \neq t$ and 0 otherwise.

A distance function was then learned using these initial sets of targets and impostors. For each class 2 targets and 150 impostors per training window were used. Once a distance function is learned, it can be used to recompute the sets of targets and impostors for learning a new distance function, and iterate this learning process. The table below shows the *area under the* receiver operating characteristic *curve* (AUC) for the first 10 iterations of distance learning, on both the training and testing examples. There is a steady increase in AUC on training examples for the first four iterations, with only a slight improvement in testing AUC; after the fifth iteration, no further improvement is seen.

Iteration	1	2	3	4	5	6	7	8	9	10
training AUC	86.3	93.9	98.9	99.3	99.3	99.4	99.3	99.3	99.3	99.4
testing AUC	83.8	82.5	84.9	84.8	85	84.8	84.6	84.6	84.6	84.3

This same training procedure was also used using *random examples* from the correct and incorrect class databases for the targets and imposters. Using random targets and impostors, the training and testing AUC values were respectively 85.8 and 88.7 after a single iteration. While distance learning is effective, it is overfitting to the training data, most likely due to the small number of training examples used. Increasing the set of training examples led to prohibitively long running times for solving the linear program to find the optimal distance. Consequently, the distance functions learned on random points are used in the experiments that apply predicted coreness to improve the Facet estimator, as they generalize better.

9.4.1.2   Mapping Distance to Coreness

Figure 9.1 shows on its vertical axis the average true coreness of examples, superimposed with the fitted logistic transform function for predicted coreness, and on its horizontal axis the corresponding 1-nearest-neighbor distance, for one training fold of examples. The blue and red lines show the average coreness of the examples in the training set for which the nearest neighbor is in a core class and a structured non-core class, respectively. The top and bottom green curves show the two logistic transform functions for the core and non-core classes, respectively, fitted to this training data (which are used when predicting column coreness on testing data). Clearly the green logistic curves fit the data quite well. Note the steep transition from high to low coreness when the nearest neighbor is from a core class.

9.4.1.3   Estimating the Number of Core Columns

For function $L(S)$ that estimates the number of core columns in the unknown reference alignment, the linear programming approach described in Sect. 9.3.1.1 to find optimal coefficients gave the fitted estimator,

$$(1.020)\, \ell_{\min} p_{\max}\, r_{\mathrm{mm}} \;+\; (0.151)\, \ell_{\min}\, r_{\mathrm{mm}} \;+$$
$$(0.035)\, \ell_{\mathrm{avg}} p_{\max}\, r_{\mathrm{am}} \;+\; (0.032)\, \ell_{\mathrm{avg}} p_{\min}\, r_{\min} \;+$$
$$(0.003)\, \ell_{\max} p_{\mathrm{avg}}\, r_{\mathrm{avg}} \,.$$

Fig. 9.2 Correlation of the estimated number and true number of core columns.

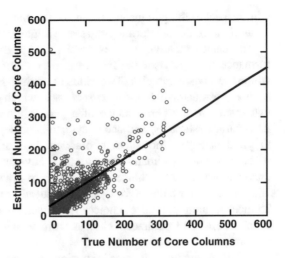

Figure 9.2 shows the correlation between the estimated number of core columns and the true number of core columns for each benchmark. The fitted estimator correlates well with the true number of core columns, but tends to overestimate, possibly due to larger benchmarks having columns that are very close to being core.

9.4.2   Improving Parameter Advising

Recall from earlier chapters that the task of parameter advising is to select a choice of values for the parameters of the alignment scoring function for a multiple sequence alignment tool, based on the set of input sequences to align. A *parameter advisor* has two ingredients: (1) an *accuracy estimator*, which estimates the accuracy of a computed alignment (for which the reference is unavailable); and (2) an *advisor set*, which is the set of assignments of values to the aligner's parameters that are considered by the advisor. The advisor picks the choice of values from the advisor set for which the aligner produces a computed alignment of the input sequences of highest estimated accuracy. In the experiments that follow, the performance of parameter advising is assessed using the Facet accuracy estimator modified by predicted coreness. For comparison, the advising accuracy of the TCS estimator, an unmodified version of Facet, and three versions of Facet modified using the column quality scores of TCS, ZORRO, and true coreness are also assessed. Modifying the estimator using true coreness is used to show a theoretical upper bound on the improvement possible if we could predict coreness perfectly.

This study is focused on parameter advising for the multiple sequence alignment tool Opal [106, 107]. While parameter advising increases the accuracy of many of the popular alignment tools (see Chap. 7), Opal is an ideal test bed for parameter advising, as in contrast to other aligners, it computes subalignments that are *optimal* with respect to the parameter choice for the sum-of-pairs scoring function at each node of the guide tree during progressive alignment.

The choice of advisor set is crucial for parameter advising. Clearly the performance of an advisor is limited by the quality of the parameter settings from which the advisor can pick. Two kinds of advisor sets are considered (see Chap. 5): accuracy-estimator-independent *oracle sets*, which contain an optimal set of choices that maximize the performance of a perfect advisor that uses true accuracy for its accuracy estimator; and accuracy-estimator-dependent *greedy sets*, which tend to yield better performance in practice than oracle sets, but are tuned for a specific accuracy estimator. Finding such advisor sets requires specifying a finite universe of parameter choices from which to draw the set. Starting from roughly 16,900 parameter choices for Opal, a reduced universe is formed by selecting the 25 most accurate parameter choices from each benchmark difficulty bin. This gave a universe of 243 parameter choices from which to construct oracle and greedy advisor sets.

When evaluating average advising accuracy on benchmarks, the over-representation of easy-to-align benchmarks is corrected by weighting benchmarks according to the same hardness bins described earlier. The weight of a benchmark falling in bin b is $(1/10)(1/n_b)$, where n_b is the number of benchmarks in bin b. These weights are such that each hardness *bin* contributes equally to the advising accuracy, which effectively uniformly averages advising accuracy across the full range of hardnesses.

Note that under this equal weighting of hardness bins, an advisor that uses only the single optimal default parameter choice will have an average advising accuracy of roughly 50% (illustrated later in Fig. 9.3). This establishes as a point of reference an average advising accuracy of 50% as the *baseline* against which to compare advising performance.

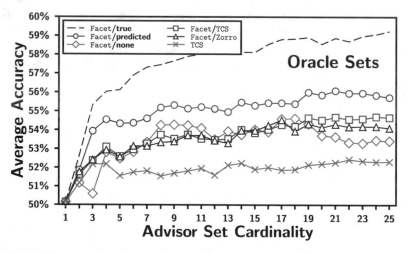

Fig. 9.3 Advising accuracy using oracle sets with the modified Facet or TCS estimators.

Note that if instead advising accuracy was measured by uniformly averaging over *benchmarks*, then the predominance of easy benchmarks (for which little improvement is possible over the default parameter choice) makes both good and bad advisors tend to an average accuracy of nearly 100%. Uniformly averaging over *bins* allows for easier discrimination among advisors, though a typical value for average advising accuracy is now pulled down from 100% toward 50%.

9.4.2.1   Modifying the `Facet` Accuracy Estimator

Exploration of the new coreness estimator, as well as TCS and ZORRO, to modify the existing features of `Facet` according to the procedure described in Sect. 9.3.2, and also the new Predicted Alignment Coreness feature described in Sect. 9.3.1, is undertaken in the context of parameter advising. For the existing feature functions that can be modified by coreness, using both the original and modified features are considered. True coreness (as opposed to predicted coreness) is also explored, which provides a theoretical limit on what is possible with a perfect coreness estimator. Coefficients are learned for the feature functions of all these variants of `Facet` separately, using the difference-fitting technique described in Chap. 2.

The new alignment accuracy estimator that uses our coreness estimator has non-zero coefficients for seven features: our new feature, Predicted Alignment Coreness F_{AC}; two features that have been modified with predicted coreness, namely, Amino Acid Identity F'_{AI} and Secondary Structure Identity F'_{SI}; and the four original features Gap Open Density F_{GO}, Secondary Structure Agreement F_{SA}, Amino Acid Identity F_{AI}, and Secondary Structure Blockiness F_{BL}. The fitted accuracy estimator that uses predicted coreness is,

$$(0.512)\,F_{GO}\;+\;(0.304)\,F'_{SI}\;+\;(0.157)\,F_{SA}\;+\;(0.109)\,F_{AI}\;+$$
$$(0.096)\,F_{BL}\;+\;(0.025)\,F'_{AI}\;+\;(0.013)\,F_{AC}\,.$$

These feature functions have different ranges, so the magnitudes of the coefficients do not necessarily correspond to the importance of the features.

9.4.2.2   Improvement on Oracle Advisor Sets

Figure 9.3 compares these various accuracy estimators in the context of parameter advising using estimator-independent *oracle* advisor sets (see Chap. 5). The horizontal axis is the cardinality of the advisor set, i.e. the number of parameter choices available to the advisor, while the vertical axis is average advising accuracy using various accuracy estimators. The advising accuracy is compared using different versions of `Facet`, as well as using the TCS accuracy estimator, on the same oracle sets, to isolate the effect each modification to the accuracy estimator has on advising performance. Using our new coreness predictor to modify the features of `Facet` increases the accuracy of parameter advising by as much as much as 3%, compared

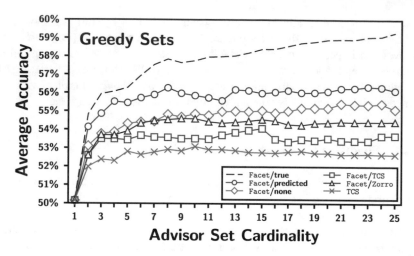

Fig. 9.4 Advising accuracy using greedy sets with the modified Facet or TCS estimators.

to the original unmodified version. This increase is in addition to the improvement of unmodified Facet over TCS, the next-best accuracy estimator in the literature.

9.4.2.3 Improvement on Greedy Advisor Sets

The results from the preceding section show the effect of using different accuracy estimators on the same advisor sets of parameter choices. Here the effect of using different accuracy estimators are shown on *greedy* advisor sets (see Chap. 5), which are near-optimal accuracy-estimator-dependent advisor sets that are designed to boost the advising accuracy when using a given accuracy estimator.

Figure 9.4 shows the advising accuracy using the Facet estimator modified by true coreness, predicted coreness, TCS, and ZORRO, using greedy advisor sets found specifically for each of these accuracy estimators. (Here each accuracy estimator is used with a different advisor set learned specifically for it by a greedy algorithm.) Once again the horizontal axis is the advisor set cardinality, and the vertical axis is advising accuracy averaged over the testing benchmarks in all folds. Using the new coreness predictor boosts the advising accuracy over using the original estimator by almost 2% when using a greedy advisor set of cardinality 7. In contrast, using TCS and ZORRO to modify features actually reduces the advising accuracy of greedy sets.

Figure 9.5 shows the advising accuracy on both training and testing benchmarks for the Facet estimator modified by predicted coreness using greedy advisor sets. The drop between training and testing accuracy suggests that by improving the generalization of greedy sets, further improvement in advising accuracy should be possible.

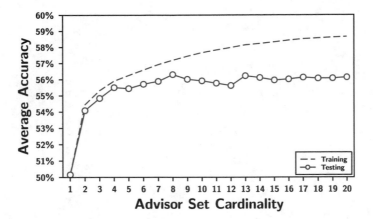

Fig. 9.5 Training and testing advising accuracy using Facet with predicted coreness.

Summary

This chapter developed a column coreness estimator for protein multiple sequence alignments that uses a regression function on nearest neighbor distances for class distance functions learned by solving a new linear programming formulation. When applied to alignment accuracy estimation and parameter advising, the coreness estimator strongly outperforms others from the literature, and gives a significant boost in accuracy for advising.

Chapter 10
Future Directions

This monograph has addressed a major task in protein multiple sequence alignment: how to automatically choose a good parameter setting for the multitude of tunable parameters in an aligner. The chapters have both explored the theoretical underpinnings of parameter advising for multiple sequence alignment, and demonstrated its utility in practice. All of the software and datasets discussed throughout are available for use by other researchers. The Facet software release includes a stand-alone tool for parameter advising, as well as the central components for ensemble alignment. The latest release of the Opal aligner also incorporates both parameter advising and adaptive local realignment. These implementations enable practitioners who use protein multiple sequence alignment to automatically increase the accuracy of their computed alignments without having to manually search for parameters that work well on their datasets. Links to this freely-available software are at the Facet website [24]: facet.cs.arizona.edu.

We conclude with some future research directions for extending and enhancing parameter advising.

10.1   Extending Parameter Advising

The techniques we have developed for accuracy estimation and parameter advising are applicable not only to multiple sequence alignment, but to *any* problem domain that meets a few basic requirements: namely, that it contains,

(a) a tool or set of tools that have parameters whose values affect the quality of the resulting output,

(b) benchmark examples consisting of an input and an associated ("ground truth") reference output with which to compare the quality of the resulting outputs of the tools, and

© Springer International Publishing AG 2017

D. DeBlasio, J. Kececioglu, *Parameter Advising for Multiple Sequence Alignment,*
Computational Biology 26, https://doi.org/10.1007/978-3-319-64918-4_10

(c) domain knowledge to enable discovery of feature functions that correlate
with output quality and that can be combined into a quality estimator.

Many domains satisfy these properties, and we list a few.

10.1.1 DNA and RNA Alignment

A natural way to immediately expand the scope of parameter advising is to
extend Facet beyond protein sequences. Extending the estimator to DNA and
RNA sequence alignments would enable advising for a much broader range of
users of multiple sequence alignment. This would also allow advising for other
bioinformatics tasks such as sequence assembly and whole genome alignment.
Successfully extending Facet to DNA and RNA alignments will require feature
functions that go beyond just capturing basic sequence information. As we have
shown for proteins, the features that have the strongest correlation with alignment
accuracy exploit auxiliary information like predicted secondary structure.

RNA feature functions could also use predicted secondary structure, though
this only applies to non-coding RNA molecules (ncRNAs), which form tertiary
structure without coding for proteins. RNA secondary structure can be predicted
for a new input sequence, and this has been used to modify RNA alignment
objective functions (see [23, 31, 32]); such approaches, however, typically cannot
handle pseudoknots (non-nested secondary structure pairings) when computing
an alignment by dynamic programming. Since feature functions are evaluated
on complete alignments, we can accommodate features that take into account
pseudoknots.

DNA feature functions no longer have secondary structure to exploit, but other
surrogates could guide us in identifying high-accuracy alignments. One such
proxy is the predicted category of each sequence region (which is essentially
what secondary structure predictions provide). Such category labels might include
protein-coding regions, translation start sites, potential ncRNAs, and so on. Another
labeling that could help in estimating alignment accuracy might be predicted chromatin placement. (DNA stored in chromosomes is wrapped around large proteins
called chromatin, and only small regions between chromatin are accessible.) Recent
work has shown that chromatin placement can be crucial for translational regulation,
so such locations might be conserved in high-accuracy alignments.

With DNA alignment, besides creating new feature functions for DNA
sequences, a major challenge is the lack of multiple sequence alignment
benchmarks. Without known ground-truth DNA alignments, we cannot learn
advisor sets and estimator coefficients. A recourse might be to generate benchmarks
from simulated multiple sequence alignments. In a simulation, we know the true
evolutionary history of the sequences, so we can recover the ground-truth alignment.
Simulation, however, will limit us to learning the evolutionary parameter values of
the simulation, and we may not learn the true biological parameter values that yield
biologically realistic alignments.

An accuracy estimator for DNA sequence alignments will be useful not only for global advising of DNA multiple alignments, but also for local advising of whole genome alignments through adaptive local realignment. Each section of a genome can evolve differently, and may even be rearranged or transposed from another sequence. Applying adaptive local realignment to whole genome alignments could overcome the challenge posed by heterogeneity in genomic sequences.

10.1.2 Sequence Assembly and Read Mapping

De novo sequence assembly suffers from many of the same complications as multiple sequence alignment: a multitude of tools that can be used to assemble sequences, with each tool having many parameters that can affect the output of the assembler, and no ideal way to rank assemblies obtained by different methods. With sequence assembly, there are again two basic issues. What feature functions correlate with assembly accuracy? And how do we obtain the ground-truth assembly? Standard measures of assembly quality are the N50 score, which is the *length* of the smallest contig (a contiguous layout of sequencing reads) that when placed in a listing of contigs sorted by length, covers half the expected length of the genome; and the L50 score, which is the minimum *number* of contigs that cover half the genome. We could again devise new feature functions that measure the consistency of labels on sequencing reads (like those described earlier for DNA alignments). For the ground-truth assembly, one possibility is to employ read mapping against a known reference genome, and then remove the reference to leave a result resembling an assembly. Alternatively, one could simulate a set of reads and take their known relation to the original underlying sequence as the ground truth.

Another potential application of parameter advising is to read mapping: recovering how sequencing reads align to a known reference genome. Feature functions for read mapping might differ from those for de novo assembly, since we know the reference genome. A common quality measure is the fraction of reads successfully mapped, where reads can fail to be mapped due to sequencing error or a poor choice of mapping parameters. Just as with DNA alignment and de novo assembly, there may be a lack of benchmarks, but simulations could provide ground-truth read-mapping results.

10.2 Enhancing Parameter Advising

Besides extending parameter advising to new applications, a few key issues still need to be addressed. One is the *generalization* (in the sense of machine learning) of our greedy estimator-aware advisor sets. Chapter 6 shows that greedy sets tend to not generalize well, and furthermore, exact sets generalize even worse. This behavior is exacerbated in the context of ensemble alignment in Chap. 7, where we compare default aligner advising to general aligner advising. Since the universe of parameter choices for general advising is a superset of the universe for default advising, as the size of the universe increases, advisor accuracy should also tend to increase (assuming we use a good estimator). In our experiments, this holds for training data (see Fig. 7.8), but not for testing data (see Fig. 7.9).

A possible strategy for overcoming this generalization issue with learning advisor sets might be to leverage *inverse parametric alignment* [60]. Inverse parametric alignment finds the optimal choice of values for real-valued parameters that essentially yields the highest alignment accuracy averaged across a set of benchmarks. In contrast, the algorithms we presented for finding advisor sets rely on having a fixed finite universe of parameter choices. This universe is intended to span the entire range of possible values for the tunable parameters of an aligner. While many tunable parameters are continuous, we are forced to discretize the range of possible settings to generate a finite universe. As the granularity of this discretization becomes finer, the chances of overfitting increase, assuming we keep the set of alignment benchmarks the same. Rather than relying on a fixed discrete universe, we can instead use inverse parametric alignment on a continuous universe in a greedy fashion to form an advisor set. Start by invoking inverse parametric alignment to find a single parameter choice P_1 that is optimal on average for all the available examples. Then remove any example that when aligned using P_1 already has high accuracy (which essentially removes the examples in the highest-accuracy bins discussed in earlier chapters). Over the remaining examples, which have low accuracy when aligned using P_1, again invoke inverse parametric alignment to find the optimal parameter choice P_2. We continue this procedure until either we have reached a specified number of parameter choices, or there are no longer any low-accuracy examples. Our advisor set $\mathcal{P}$ is then the parameter choices encountered,

$$\mathcal{P} = \{P_1, P_2, \ldots, P_k\}.$$

With this approach, the accuracy of the learned advisor is no longer limited by a predefined parameter universe.

We also have a similar generalization issue in core column prediction. While considerable work was done in Chap. 9 to learn good distance functions on training data for nearest neighbor search in our approach to coreness prediction, the distance functions learned were limited by the available computational resources. With more computing time, we may be able to learn better distance functions that yield a more accurate coreness predictor with the methods already developed. Additionally, we might apply other machine learning techniques for distance metric learning, including kernel transformations on the input data to reduce its dimensionality.

Clearly there are many promising directions to take this new methodology of parameter advising.

References

1. Ahola, V., Aittokallio, T., Vihinen, M., Uusipaikka, E.: A statistical score for assessing the quality of multiple sequence alignments. BMC Bioinform. **7**(484), 1–19 (2006)
2. Ahola, V., Aittokallio, T., Vihinen, M., Uusipaikka, E.: Model-based prediction of sequence alignment quality. Bioinformatics **24**(19), 2165–2171 (2008)
3. Altschul, S.F., Gish, W., Miller, W., Myers, E.W., Lipman, D.J.: Basic local alignment search tool. J. Mol. Biol. **215**(3), 403–410 (1990)
4. Aniba, M.R., Poch, O., Marchler-Bauer, A., Thompson, J.D.: AlexSys: a knowledge-based expert system for multiple sequence alignment construction and analysis. Nucleic Acids Res. **38**(19), 6338–6349 (2010)
5. Anson, E.L., Myers, E.W.: ReAligner: a program for refining DNA sequence multi-alignments. J. Comput. Biol. **4**(3), 369–83 (1997)
6. Apweiler, R., Bairoch, A., Wu, C.H., Barker, W.C., Boeckmann, B., Ferro, S., Gasteiger, E., Huang, H., Lopez, R., Magrane, M., Martin, M.J., Natale, D.A., O'Donovan, C., Redaschi, N., Yeh, L.S.L.: UniProt: the Universal Protein knowledgebase. Nucleic Acids Res. **32**(Database), D115–D119 (2004)
7. Armougom, F., Moretti, S., Keduas, V., Notredame, C.: The iRMSD: a local measure of sequence alignment accuracy using structural information. Bioinformatics **22**, E35–E39 (2006)
8. Bahr, A., Thompson, J.D., Thierry, J.C., Poch, O.: BAliBASE (Benchmark Alignment dataBASE): enhancements for repeats, transmembrane sequences and circular permutations. Nucleic Acids Res. **29**(1), 323–326 (2001)
9. Balaji, S., Sujatha, S., Kumar, S.S.C., Srinivasan, N.: PALI: a database of Phylogeny and ALIgnment of homologous protein structures. Nucleic Acids Res. **29**(1), 61–65 (2001)
10. Berman, H.M., Westbrook, J., Feng, Z., Gilliland, G., Bhat, T.N., Weissig, H., Shindyalov, I.N., Bourne, P.E.: The Protein Data Bank. Nucleic Acids Res. **28**(1), 35–242 (2000)
11. Beygelzimer, A., Kakade, S., Langford, J.: Cover trees for nearest neighbor. In: Proceedings of the 23rd International Conference on Machine Learning (ICML), pp. 97–104 (2006)
12. Bradley, R.K., Roberts, A., Smoot, M., Juvekar, S., Do, J., Dewey, C., Holmes, I., Pachter, L.: Fast statistical alignment. PLoS Comput. Biol. **5**(5), 1–15 (2009)
13. Bucka-Lassen, K., Caprani, O., Hein, J.: Combining many multiple alignments in one improved alignment. Bioinformatics **15**(2), 122–130 (1999)
14. Camon, E., Magrane, M., Barrell, D., Lee, V., Dimmer, E., Maslen, J., Binns, D., Harte, N., Lopez, R., Apweiler, R.: The Gene Ontology Annotation (GOA) Database: sharing knowledge in Uniprot with Gene Ontology. Nucleic Acids Res. **32**(90001), 262D–266 (2004)

© Springer International Publishing AG 2017

D. DeBlasio, J. Kececioglu, *Parameter Advising for Multiple Sequence Alignment*,
Computational Biology 26, https://doi.org/10.1007/978-3-319-64918-4

15. Capella-Gutierrez, S., Silla-Martinez, J.M., Gabaldón, T.: trimAl: a tool for automated alignment trimming in large-scale phylogenetic analyses. Bioinformatics 25(15), 1972–1973 (2009)
16. Carrillo, H., Lipman, D.: The multiple sequence alignment problem in biology. SIAM J. Appl. Math. 48(5), 1073–1082 (1988)
17. Castresana, J.: Selection of conserved blocks from multiple alignments for their use in phylogenetic analysis. Mol. Biol. Evol. 17(4), 540–552 (2000)
18. Chang, J.M., Tommaso, P.D., Notredame, C.: TCS: a new multiple sequence alignment reliability measure to estimate alignment accuracy and improve phylogenetic tree reconstruction. Mol. Biol. Evol. 31(6), 1625–1637 (2014)
19. Collingridge, P.W., Kelly, S.: MergeAlign: improving multiple sequence alignment performance by dynamic reconstruction of consensus multiple sequence alignments. BMC Bioinform. 13(117), 1–10 (2012)
20. Cormen, T.H., Leiserson, C.E., Rivest, R.L., Stein, C.: Introduction to Algorithms, 3rd edn. MIT Press, Cambridge (2009)
21. Darling, A.C., Mau, B., Blattner, F.R., Perna, N.T.: Mauve: multiple alignment of conserved genomic sequence with rearrangements. Genome Res. 14(7), 1394–1403 (2004)
22. Dayhoff, M.O., Schwartz, R.M., Orcutt, B.C.: A model of evolutionary change in proteins. In: Atlas of Protein Sequences and Structure, vol. 5, pp. 345–352. National Biomedical Research Foundation, Silver Spring (1978)
23. DeBlasio, D.F.: New Computational Approaches for Multiple RNA Alignment and RNA Search. Masters Thesis. University of Central Florida, Orlando, Florida (2009)
24. DeBlasio, D., Kececioglu, J.: Learning parameter sets for alignment advising. In: Proceedings of the 5th ACM Conference on Bioinformatics, Computational Biology, and Health Informatics (ACM-BCB), pp. 230–239 (2014)
25. DeBlasio, D.F., Kececioglu, J.D.: Facet: software for accuracy estimation of protein multiple sequence alignments (version 1.1) (2014). http://facet.cs.arizona.edu
26. DeBlasio, D., Kececioglu, J.: Ensemble multiple sequence alignment via advising. In: Proceedings of the 6th ACM Conference on Bioinformatics, Computational Biology, and Health Informatics (ACM-BCB), pp. 452–461 (2015)
27. DeBlasio, D., Kececioglu, J.D.: Predicting core columns of protein multiple sequence alignments for improved parameter advising. In: Proceedings of the 16th Workshop on Algorithms in Bioinformatics (WABI), pp. 77–89 (2016)
28. DeBlasio, D.F., Kececioglu, J.D.: Learning parameter-advising sets for multiple sequence alignment. IEEE/ACM Trans. Comput. Biol. Bioinform. 14(5), 1028–1041 (2017)
29. DeBlasio, D., Kececioglu, J.D.: Core column prediction for protein multiple sequence alignments. Algorithms Mol. Biol. 12, 1–16 (2017)
30. DeBlasio, D., Kececioglu, J.D.: Estimating the accuracy of multiple alignments and its use in parameter advising. In: Proceedings of the 21st Conference on Research in Computational Molecular Biology (RECOMB), pp. 1–17 (2017)
31. DeBlasio, D., Bruand, J., Zhang, S.: PMFastR: a new approach to multiple RNA structure alignment. In: Proceedings of the 9th International Conference on Algorithms in Bioinformatics (WABI'09), pp. 49–61 (2009)
32. DeBlasio, D., Bruand, J., Zhang, S.: A memory efficient method for structure-based RNA multiple alignment. IEEE/ACM Trans. Comput. Biol. Bioinform. 9(1), 1–11 (2012)
33. DeBlasio, D.F., Wheeler, T.J., Kececioglu, J.D.: Estimating the accuracy of multiple alignments and its use in parameter advising. In: Proceedings of the 16th Conference on Research in Computational Molecular Biology (RECOMB), pp. 45–59 (2012)
34. Do, C.B., Mahabhashyam, M.S.P., Brudno, M., Batzoglou, S.: ProbCons: probabilistic consistency-based multiple sequence alignment. Genome Res. 15(2), 330–340 (2005)
35. Dress, A.W., Flamm, C., Fritzsch, G., Grünewald, S., Kruspe, M., Prohaska, S.J., Stadler, P.F.: Noisy: identification of problematic columns in multiple sequence alignments. Algorithms Mol. Biol. 3(7), 1–10 (2008)

36. Durbin, R., Eddy, S.R., Krogh, A., Mitchison, G.: Biological Sequence Analysis: Probabilistic Models of Proteins and Nucleic Acids. Cambridge University Press, Cambridge (1998)
37. Edgar, R.C.: MUSCLE: multiple sequence alignment with high accuracy and high throughput. Nucleic Acids Res. 32(5), 1792–1797 (2004)
38. Edgar, R.C.: MUSCLE: a multiple sequence alignment method with reduced time and space complexity. BMC Bioinform. 5(113), 1–19 (2004)
39. Edgar, R.C.: BENCH (2009). http://www.drive5.com/bench
40. Estabrook, G., Johnson, C., Morris, F.M.: An idealized concept of the true cladistic character. Math. Biosci. 23(3), 263–272 (1975)
41. Feng, D.F., Doolittle, R.F.: Progressive sequence alignment as a prerequisite to correct phylogenetic trees. J. Mol. Evol. 25(4), 351–360 (1987)
42. Finn, R.D., Mistry, J., Tate, J., Coggill, P., Heger, A., Pollington, J.E., Gavin, O.L., Gunasekaran, P., Ceric, G., Forslund, K., Holm, L., Sonnhammer, E.L.L., Eddy, S.R., Bateman, A.: The Pfam protein families database. Nucleic Acids Res. 38(Database), D211–D222 (2009)
43. Fitch, W.M., Margoliash, E.: A method for estimating the number of invariant amino acid coding positions in a gene using cytochrome c as a model case. Biochem. Genet. 1(1), 65–71 (1967)
44. Garey, M.R., Johnson, D.S.: Computers and Intractability: A Guide to the Theory of NP-Completeness. W.H. Freeman and Company, New York (1979)
45. Gotoh, O.: An improved algorithm for matching biological sequences. J. Mol. Biol. 162(3), 705–508 (1982)
46. Gotoh, O.: Optimal alignment between groups of sequences and its application to multiple sequence alignment. Comput. Appl. Biosci. 9(3), 361–370 (1993)
47. Gusfield, D.: Algorithms on Strings, Trees, and Sequences: Computer Science and Computational Biology. Cambridge University Press, New York, NY (1997)
48. Henikoff, S., Henikoff, J.G.: Amino acid substitution matrices from protein blocks. Proc. Natl. Acad. Sci. U. S. A. 89(22), 10915–10919 (1992)
49. Hertz, G.Z., Stormo, G.D.: Identifying DNA and protein patterns with statistically significant alignments of multiple sequences. Bioinformatics 15(7–8), 563–577 (1999)
50. IBM Corporation: CPLEX: High-performance mathematical programming solver for linear programming, mixed integer programming, and quadratic programming (version 12.6.2.0) (2015). http://www.ilog.com/products/cplex
51. Jones, D.T.: Protein secondary structure prediction based on position-specific scoring matrices. J. Mol. Biol. 292(2), 195–202 (1999)
52. Jones, E., Oliphant, T., Peterson, P., et al.: SciPy: open source scientific tools for Python (2001). http://www.scipy.org/
53. Karlin, S., Altschul, S.F.: Methods for assessing the statistical significance of molecular sequence features by using general scoring schemes. Proc. Natl. Acad. Sci. U. S. A. 87(6), 2264–2268 (1990)
54. Katoh, K., Misawa, K., Kuma, K.i., Miyata, T.: Maft: a novel method for rapid multiple sequence alignment based on fast fourier transform. Nucleic Acids Res. 30(14), 3059–3066 (2002)
55. Katoh, K., Kuma, K.i., Toh, H., Miyata, T.: Mafft version 5: improvement in accuracy of multiple sequence alignment. Nucleic Acids Res. 33(2), 511–518 (2005)
56. Kececioglu, J., DeBlasio, D.: Accuracy estimation and parameter advising for protein multiple sequence alignment. J. Comput. Biol. 20(4), 259–279 (2013)
57. Kececioglu, J., Kim, E.: Simple and fast inverse alignment. In: Proceedings of the 10th Conference on Research in Computational Molecular Biology (RECOMB), pp. 441–455 (2006)
58. Kececioglu, J., Starrett, D.: Aligning alignments exactly. In: Proceedings of the 8th Conference on Research in Computational Molecular Biology (RECOMB), pp. 85–96. ACM (2004)
59. Kemena, C., Taly, J.F., Kleinjung, J., Notredame, C.: STRIKE: evaluation of protein MSAs using a single 3D structure. Bioinformatics 27(24), 3385–3391 (2011)

60. Kim, E., Kececioglu, J.: Learning scoring schemes for sequence alignment from partial examples. IEEE/ACM Trans. Comput. Biol. Bioinform. **5**(4), 546–556 (2008)
61. Kim, J., Ma, J.: PSAR: measuring multiple sequence alignment reliability by probabilistic sampling. Nucleic Acids Res. **39**(15), 6359–6368 (2011)
62. Kück, P., Meusemann, K., Dambach, J., Thormann, B., von Reumont, B.M., Wägele, J.W., Misof, B.: Parametric and non-parametric masking of randomness in sequence alignments can be improved and leads to better resolved trees. Front. Zool. **7**(10), 1–12 (2010)
63. Kuznetsov, I.B.: Protein sequence alignment with family-specific amino acid similarity matrices. BMC Res. Notes **4**(296), 1–10 (2011)
64. Landan, G., Graur, D.: Heads or tails: a simple reliability check for multiple sequence alignments. Mol. Biol. Evol. **24**(6), 1380–1383 (2007)
65. Larkin, M.A., et al.: ClustalW and ClustalX version 2.0. Bioinformatics **23**(21), 2947–2948 (2007)
66. Lassmann, T., Sonnhammer, E.: Kalign: an accurate and fast multiple sequence alignment algorithm. BMC Bioinform. **6**(298), 1–9 (2005)
67. Lassmann, T., Sonnhammer, E.L.L.: Automatic assessment of alignment quality. Nucleic Acids Res. **33**(22), 7120–7128 (2005)
68. Lee, C., Grasso, C., Sharlow, M.F.: Multiple sequence alignment using partial order graphs. Bioinformatics **18**(3), 452–464 (2002)
69. Liu, Y., Schmidt, B., Maskell, D.L.: MSAProbs: multiple sequence alignment based on pair hidden Markov models and partition function posterior probabilities. Bioinformatics **26**(16), 1958–1964 (2010)
70. Liu, K., Warnow, T.J., Holder, M.T., Nelesen, S.M., Yu, J., Stamatakis, A.P., Linder, C.R.: SATé-II: very fast and accurate simultaneous estimation of multiple sequence alignments and phylogenetic trees. Syst. Biol. **61**(1), 90–106 (2011)
71. Loytynoja, A., Goldman, N.: An algorithm for progressive multiple alignment of sequences with insertions. Proc. Natl. Acad. Sci. U. S. A. **102**(30), 10557–10562 (2005)
72. Misof, B., Misof, K.: A Monte Carlo approach successfully identifies randomness in multiple sequence alignments: a more objective means of data exclusion. Syst. Biol. **58**(1), 21–34 (2009)
73. Moré, J.J., Sorensen, D.C., Hillstrom, K.E., Garbow, B.S.: The MINPACK project. In: Sources and Development of Mathematical Software, pp. 88–111. Prentice-Hall, Englewood Cliffs (1984)
74. Müller, T., Spang, R., Vingron, M.: Estimating amino acid substitution models: a comparison of Dayhoff's estimator, the resolvent approach and a maximum likelihood method. Mol. Biol. Evol. **19**(1), 8–13 (2002)
75. Muller, J., Creevey, C.J., Thompson, J.D., Arendt, D., Bork, P.: AQUA: automated quality improvement for multiple sequence alignments. Bioinformatics **26**(2), 263–265 (2010)
76. Needleman, S.B., Wunsch, C.D.: A general method applicable to the search for similarities in the amino acid sequence of two proteins. J. Mol. Biol. **48**(3), 443–453 (1970)
77. Notredame, C., Holm, L., Higgins, D.G.: COFFEE: an objective function for multiple sequence alignments. Bioinformatics **14**(5), 407–422 (1998)
78. Notredame, C., Higgins, D.G., Heringa, J.: T-Coffee: a novel method for fast and accurate multiple sequence alignment. J. Mol. Biol. **302**(1), 205–217 (2000)
79. Ortuño, F.M., Valenzuela, O., Pomares, H., Rojas, F., Florido, J.P., Urquiza, J.M., Rojas, I.: Predicting the accuracy of multiple sequence alignment algorithms by using computational intelligent techniques. Nucleic Acids Res. **41**(1), e26–e26 (2012)
80. Ortuño, F., Valenzuela, O., Pomares, H.e., Rojas, I.: Evaluating multiple sequence alignments using a LS-SVM approach with a heterogeneous set of biological features. In: Proceedings of the 12th International Work-Conference on Artificial Neural Networks (IWANN 2013), pp. 150–158 (2013)
81. Pei, J., Grishin, N.V.: AL2CO: calculation of positional conservation in a protein sequence alignment. Bioinformatics **17**(8), 700–712 (2001)

82. Pei, J., Grishin, N.V.: MUMMALS: multiple sequence alignment improved by using hidden Markov models with local structural information. Nucleic Acids Res. **34**(16), 4364–4374 (2006)

83. Pei, J., Grishin, N.V.: PROMALS: towards accurate multiple sequence alignments of distantly related proteins. Bioinformatics **23**(7), 802–808 (2007)

84. Pei, J., Sadreyev, R., Grishin, N.V.: PCMA: fast and accurate multiple sequence alignment based on profile consistency. Bioinformatics **19**(3), 427–428 (2003)

85. Penn, O., Privman, E., Landan, G., Graur, D., Pupko, T.: An alignment confidence score capturing robustness to guide tree uncertainty. Mol. Biol. Evol. **27**(8), 1759–1767 (2010)

86. Prakash, A., Tompa, M.: Assessing the discordance of multiple sequence alignments. IEEE/ACM Trans. Comput. Biol. Bioinform. **6**(4), 542–551 (2009)

87. Raghava, G., Searle, S.M., Audley, P.C., Barber, J.D., Barton, G.J.: OXBench: a benchmark for evaluation of protein multiple sequence alignment accuracy. BMC Bioinform. **4**(1), 1–23 (2003)

88. Ren, J.: SVM-based automatic annotation of multiple sequence alignments. J. Comput. **9**(5), 1109–1116 (2014)

89. Roshan, U., Livesay, D.R.: PROBALIGN: multiple sequence alignment using partition function posterior probabilities. Bioinformatics **22**(22), 2715–2721 (2006)

90. Roskin, K.M., Paten, B., Haussler, D.: Meta-alignment with Crumble and Prune: partitioning very large alignment problems for performance and parallelization. BMC Bioinform. **12**(1), 1–12 (2011)

91. Schwartz, A.S., Pachter, L.: Multiple alignment by sequence annealing. Bioinformatics **23**(2), e24–e29 (2007)

92. Sela, I., Ashkenazy, H., Katoh, K., Pupko, T.: GUIDANCE2: accurate detection of unreliable alignment regions accounting for the uncertainty of multiple parameters. Nucleic Acids Res. **43**(W1), W7–W14 (2015)

93. Sievers, F., et al.: Fast, scalable generation of high-quality protein multiple sequence alignments using Clustal Omega. Mol. Syst. Biol. **7**(1), 539–539 (2011)

94. Subramanian, A.R., Weyer-Menkhoff, J., Kaufmann, M., Morgenstern, B.: DIALIGN-T: an improved algorithm for segment-based multiple sequence alignment. BMC Bioinform. **6**(66), 1–13 (2005)

95. Subramanian, A.R., Kaufmann, M., Morgenstern, B.: DIALIGN-TX: greedy and progressive approaches for segment-based multiple sequence alignment. Algorithms Mol. Biol. **3**(6), 1–11 (2008)

96. Suzek, B.E., Huang, H., McGarvey, P., Mazumder, R., Wu, C.H.: UniRef: comprehensive and non-redundant uniprot reference clusters. Bioinformatics **23**(10), 1282–1288 (2007)

97. The UniProt Consortium: the universal protein resource (uniprot). Nucleic Acids Res. **35**(suppl 1), D193–D197 (2007)

98. Thompson, J.D., Higgins, D.G., Gibson, T.J.: ClustalW: improving the sensitivity of progressive multiple sequence alignment through sequence weighting, position-specific gap penalties and weight matrix choice. Nucleic Acids Res. **22**(22), 4673–4680 (1994)

99. Thompson, J.D., Plewniak, F., Ripp, R., Thierry, J.C., Poch, O.: Towards a reliable objective function for multiple sequence alignments. J. Mol. Biol. **314**(4), 937–951 (2001)

100. Thompson, J.D., Thierry, J.C., Poch, O.: RASCAL: rapid scanning and correction of multiple sequence alignments. Bioinformatics **19**(9), 1155–1161 (2003)

101. Thompson, J.D., Prigent, V., Poch, O.: LEON: multiple aLignment Evaluation Of Neighbours. Nucleic Acids Research **32**(4), 1298–1307 (2004)

102. Van Walle, I., Lasters, I., Wyns, L.: SABmark: a benchmark for sequence alignment that covers the entire known fold space. Bioinformatics **21**(7), 1267–1268 (2005)

103. Wallace, I.M., O'Sullivan, O., Higgins, D.G., Notredame, C.: M-Coffee: combining multiple sequence alignment methods with T-Coffee. Nucleic Acids Res. **34**(6), 1692–1699 (2006)

104. Wang, L., Jiang, T.: On the complexity of multiple sequence alignment. J. Comput. Biol. J. Comput. Mol. Cell Biol. **1**(4), 337–348 (1994)

105. Weinberger, K.Q., Saul, L.K.: Distance metric learning for large margin nearest neighbor classification. J. Mach. Learn. Res. **10**, 207–244 (2009)

106. Wheeler, T.J., Kececioglu, J.D.: Multiple alignment by aligning alignments. In: Proceedings of the 15th ISCB Conference on Intelligent Systems for Molecular Biology (ISMB), Bioinformatics, vol. 23(13), pp. i559–i568 (2007)

107. Wheeler, T.J., Kececioglu, J.D.: `Opal`: software for aligning multiple biological sequences (version 2.1.0) (2012). http://opal.cs.arizona.edu

108. Wilbur, W.J., Lipman, D.J.: Rapid similarity searches of nucleic acid and protein data banks. Proc. Natl. Acad. Sci. U. S. A. **80**, 726–730 (1983)

109. Wilcoxon, F.: Individual comparisons by ranking methods. Biom. Bull. **1**(6), 80–83 (1945)

110. Will, S., Reiche, K., Hofacker, I.L., Stadler, P.F., Backofen, R.: Inferring noncoding RNA families and classes by means of genome-scale structure-based clustering. PLoS Comput. Biol. **3**(4), 680–691 (2007)

111. Woerner, A., Kececioglu, J.: Faster metric-space nearest-neighbor search using dispersion trees (2017). In preparation

112. Wu, M., Chatterji, S., Eisen, J.A.: Accounting for alignment uncertainty in phylogenomics. PLoS One **7**(1), 1–10 (2012)

113. Yang, Z.: Maximum-likelihood estimation of phylogeny from DNA sequences when substitution rates differ over sites. Mol. Biol. Evol. **10**(6), 1396–1401 (1993)

114. Ye, Y., Cheung, D.W.l., Wang, Y., Yiu, S.M., Zhang, Q., Lam, T.W., Ting, H.F.: `GLProbs`: aligning multiple sequences adaptively. IEEE/ACM Trans. Comput. Biol. Bioinform. (TCBB) **12**(1), 67–78 (2015)

115. Zhihua, Z.: Ensemble Methods: Foundations and Algorithms. Chapman and Hall, New York (2012)

Author Index

© Springer International Publishing AG 2017
D. DeBlasio, J. Kececioglu, *Parameter Advising for Multiple Sequence Alignment*,
Computational Biology 26, https://doi.org/10.1007/978-3-319-64918-4

Subject Index

A
a priori advising, 7, 10
accuracy estimator, 4, 6, 7, 19, 128
    scoring-function-based, 7
    support-based, 7–9
advisor set, 5, 53
    greedy set, 57, 72, 136
    oracle set, 56, 109, 135
aligner advising, 88, 99
aligners, 2
approximation algorithm, 57
approximation ratio, 58

B
benchmark alignment, 39, 42, 66, 108
bin packing, 31

C
column confidence scoring, 7, 12
core columns, 3, 13, 30, 38, 117
correctness, 7
cross validation, 68, 94

D
default parameter, 2, 42
dispersion tree, 122

E
example alignment, 6, 21, 42

F
Facet, 5, 29, 69, 94
feature functions, 19, 30, 130

G
GUIDANCE, 10

I
information content, 38
inverse parametric alignment, 68

L
linear programming, 22, 48, 123
    integer linear programming, 52
longest source-sink path, 32

M
M-Coffee, 11, 96
MergeAlign, 11, 96
meta-alignment, 7, 11
MOS, 9, 70
multiple sequence alignment, 1
    progressive alignment, 2

N
nearest neighbor, 122
NorMD, 8, 70
NP-complete, 1, 41

© Springer International Publishing AG 2017
D. DeBlasio, J. Kececioglu, *Parameter Advising for Multiple Sequence Alignment*,
Computational Biology 26, https://doi.org/10.1007/978-3-319-64918-4